Routledge Rev

Socio-Economic Models in Geography

First published in 1968, this book explores the theme of geographical generalization, or model building. It is composed of seven of the chapters from the original Models in Geography, published in 1967. The first chapter broadly outlines this theme and examines the nature and function of generalized statements, ranging from conceptual models to scale models, in a geographical context. The following six chapters deal with socio-economic building in geography. They focus on demographic and sociological models as well as looking at special aspects of models in human geography in reference to economic development, urban geography and settlement location, industrial location, and agricultural activity.

This book represents a robustly anti-idiographic statement of modern work in one of the major branches of geography.

Socio-Economic Models in Geography

Edited by
Richard J. Chorley and Peter Haggett

Routledge
Taylor & Francis Group

First published in 1967
by Methuen & Co Ltd
First published in University Paperback in 1968

This edition first published in 2013 by Routledge
2 Park Square, Milton Park, Abingdon, Oxon, OX14 4RN

Simultaneously published in the USA and Canada
by Routledge
711 Third Avenue, New York, NY 10017

Routledge is an imprint of the Taylor & Francis Group, an informa business

© 1967 Methuen & Co Ltd

Publisher's Note
The publisher has gone to great lengths to ensure the quality of this reprint but
points out that some imperfections in the original copies may be apparent.

Disclaimer
The publisher has made every effort to trace copyright holders and welcomes
correspondence from those they have been unable to contact.

A Library of Congress record exists under ISBN: 68103561

ISBN 13: 978-0-415-64544-7 (hbk)
ISBN 13: 978-0-203-07861-7 (ebk)
ISBN 13: 978-0-415-64545-4 (pbk)

Socio-Economic
MODELS IN GEOGRAPHY

Edited by

RICHARD J. CHORLEY

PETER HAGGETT

UNIVERSITY PAPERBACKS

METHUEN: LONDON

First published in 1967
First published as a University Paperback in 1968
© 1967 Methuen & Co Ltd
Printed in Great Britain
by Ebenezer Baylis & Son Ltd
The Trinity Press
Worcester and London
S B N 416 29630 0

Distributed in the USA
by Barnes and Noble Inc. New York

University Paperbacks are published by
METHUEN & CO LTD
11 New Fetter Lane London EC4

Contents

CHAPTER ONE

Models, Paradigms and the New Geography

P. HAGGETT and R. J. CHORLEY

Models are undeniably beautiful, and a man may justly be proud to be seen in their company. But they may have their hidden vices. The question is, after all, not only whether they are good to look at, but whether we can live happily with them.　　　　KAPLAN, 1964, p. 288.

In concluding the previous volume in this series, we attempted to review the paths taken by various workers in moving towards what they saw to be the 'frontier' in geographical research. We argued there that the quest for a model or models was a recurrent theme in their search. This volume is a direct outcome of that conclusion in that specific workers were asked to discuss the role of model-building within their own special fields of geographical research. While we would not wish to pre-judge their findings, it will be evident from the format and arrangement of the chapters that there is: (1) some measurable contrast between their approaches to geography, various as they are, and those that characterize the great part of established geographical patterns of thinking, as evidenced by existing textbooks and syllabuses; and (2) a community of common ideas that link all contributors into what Price (1963, p. 62) would characterize as an 'invisible college' of geographical practitioners. Whether this communality is sufficient to form the basis of what Manley (1966) has termed a 'New Geography' is not for us to judge. However, it is perhaps significant that the greater part of the volume is based on work produced since 1945, and much of it since 1960. In this opening chapter we discuss what we believe to be the significance of this new search for a model-based geography.

FACTS, MODELS AND PARADIGMS

The nature of facts

Information in geography is capable of treatment in terms of general information theory. In this context factual information only has relevance within some more general frame of reference, and such a basic operation as the definition of a relevant fact can only be made on the basis of some theoretical framework. There are also different levels of organization of relevant information. Some information can be relevantly organized only at a small scale, whereas the orderly large-scale patterns of other information are blurred or swamped altogether on the local scale. One can therefore view geographical information registration and analysis, from one point of view at any rate, as a problem in the separation of regional and local information patterns from the more randomly-organized information which, as 'noise', obscures them (Chorley and Haggett, 1965). Of course, one may choose to regard the noise

1.1 A photograph of melting snow taken on impulse by a photographer in China just before the last war. The pattern makes no sense until it is organized as a full-face and shoulders, similar in style to a late-medieval representation of Christ; the upper margin cutting the brow and illuminated from the right (*Source: Partly from Porter, 1954*).

as the more significant element and to ask whether it is useful to try to recognize any order in reality. This results in the stress being placed on the variety of geographical information available and in attempts to subdivide information. However, it is becoming increasingly popular to ask what kinds of order are exhibited by geographical information and on what scales of space and time each operates. In short, the 'simple' registration of facts is being recognized not only as unsatisfactory but as an impossibility. Hanson (1958, pp. 8–19) has pointed out that what is observed depends not only on the context in which a particular phenomenon is set, but in the manner in which one is prepared to view it. In the words of Sigwart: 'That there is more order in the world than appears at first sight is not discovered *till the order is looked for*' (Quoted by Hanson, 1958, p. 204). Figure 1.1 gives a striking illustration both of the close apparent relationship between order and disorder, and of the subjective approach necessary to identify what it believed to be orderly. The distinction between the idiographic and nomothetic approaches to the real world was recognized by Aristotle, although not in the terms which we currently employ, when he pointed out that poetry is more philosophical and of graver import than history because it is concerned with what is pervasive and universal, whereas history is addressed to what is special and singular (Nagel, 1961, p. 547). Today the distinction is made commonly between the 'humanities' which are primarily concerned with the unique and non-recurrent, and the 'sciences' which seek to establish general statements for repeatable events and process. Contemporary geography obviously lies athwart this apparent gulf, which must either be bridged or must lead to the dismemberment of the existing discipline. The dichotomy between the general and the particular was clearly stated by Francis Bacon in his *Maxims of the Law*; 'For there be two contrary faults and extremities in the debating and sifting of the law, which may be noted in two several manner of arguments: some argue upon general grounds, and come not near the point in question; others, without laying any foundation of a ground or difference of reason, do loosely put cases, which, though they go near the point, yet being put so scattered, prove not, but rather serve to make the law appear more doubtful than to make it more plain'. Indeed, the distinction between the idiographic and nomothetic views of geography, so strongly put by Bunge (1962), may be useful in highlighting many of the current shortcomings in the subject, but is less valuable from the more purely philosophical standpoint. Bambrough (1964, p. 100), for example, points out that all reasoning is ultimately concerned with particular cases, and that laws, rules and principles are merely devices for bringing particular cases to bear on other particular cases. 'The ideal limiting case of representation is reduplication, and a duplicate is too true to be useful. Anything that falls short of the ideal limit of reduplication is too useful to be altogether true' (Bambrough, 1964, p. 98). In short, every individual is, by definition, different, but the most significant statement which can be made about modern scholarship in general is that it has been found to be intellectually more profitable, satisfying and productive to view the phenomena of the real world in terms of their 'set characteristics', rather than to concentrate upon their individual deviations from one another.

The nature of models

The catholic view of models taken in this volume derives largely from Skilling (1964). He argued that a model can be a theory or a law or an hypothesis or a structured idea. It can be a role, a relation or an equation. It can be a synthesis of data. Most important from the geographical viewpoint, it can also include reasoning about the real world by means of translations in space (to give spatial models) or in time (to give historical models).

The need for idealization. The traditional reaction of man to the apparent complexity of the world around him has been to make for himself a simplified and intelligible picture of the world. 'He then tries to substitute this cosmos of his own for the world of experience, and thus to overcome it' (Chorafas, 1965, p. 1). The mind decomposes the real world into a series of simplified systems and thus achieves in one act 'an overview of the essential characteristics of a domain' (Apostel, 1961, p. 15). This simplification requires both sensual and intellectual creativity (Keipers, 1961, p. 132). 'The mind needs to see the system in opposition and distinction to all others; therefore the separation of the system from others is made more complete than it is in reality. The system is viewed from a certain scale; details that are too microscopical or too global are of no interest to us. Therefore they are left out. The system is known or controlled within certain limits of approximation. Therefore effects that do not reach this level of approximation are neglected. The system is studied with a certain purpose in mind; everything that does not affect this purpose is eliminated. The various features of the system need to be known as aspects of one identical whole; therefore their unity is exaggerated' (Apostel, 1961, pp. 15–16). According to this view, reality exists as a patterned and bounded connexity which has been explored by the use of simplified patterns of symbols, rules and processes (Meadows, 1957, pp. 3–4). The simplified statements of this structural interdependence have been termed 'models'. A model is thus a simplified structuring of reality which presents supposedly significant features or relationships in a generalized form. Models are highly subjective approximations in that they do not include all associated observations or measurements, but as such they are valuable in obscuring incidental detail and in allowing fundamental aspects of reality to appear. This selectivity means that models have varying degrees of probability and a limited range of conditions over which they apply. The most successful models possess a high probability of application and a wide range of conditions in which they seem appropriate. Indeed, the value of a model is often directly related to its level of abstraction. However, all models are constantly in need of improvement as new information or vistas of reality appear, and the more successfully the model was originally structured the more likely it seems that such improvement must involve the construction of a different model.

Characteristics of models. The term 'model' is conventionally employed in a number of different ways. It is used as a noun implying a representation, as an adjective implying a degree of perfection, or as a verb implying to demonstrate or to show what something is like (Ackoff, Gupta and Minas, 1962, p. 108). In fact models possess all these properties.

The most fundamental feature of models is that their construction has involved a highly *selective* attitude to information, wherein not only noise but less important signals have been eliminated to enable one to see something of the heart of things. Models can be viewed as selective approximations which, by the elimination of incidental detail, allow some fundamental, relevant or interesting aspects of the real world to appear in some generalized form. Thus models can be thought of as selective pictures and 'a direct description of the logical characteristics of our knowledge of the external world shows that each of these pictures gives undue prominence to some features of our knowledge and obscures and distorts the other features that rival pictures emphasize. Each of them directs such a bright light on one part of the scene that it obscures other parts in a dark shadow' (Bambrough, 1964, p. 102). As Black (1962, p. 220) wrote of scale models, '. . . only by being unfaithful in *some* respect can a model represent its original'.

Another important model characteristic is that models are *structured*, in the sense that the selected significant aspects of the 'web of reality' are exploited in terms of their connections. It is interesting that what is often termed a model by logicians is called by econometricians a 'structure' (Suppes, 1961, p. 165; Kaplan, 1964, p. 267). Science has profited greatly from this *pattern seeking*, in which phenomena are viewed in terms of a kind of organic relationship.

This model feature leads immediately to the *suggestive* nature of models, in that a successful model contains suggestions for its own extension and generalization (Hesse, 1953–54, pp. 213–214). This implies, firstly, that the whole model structure has greater implications than a study of its individual parts might lead one to suppose (Deutsch, 1948–49), and, secondly, that predictions can be made about the real world from the model. Models have thus been termed 'speculative instruments', and Black (1962, pp. 232–233) has described a promising model as 'one with implications rich enough to suggest novel hypotheses and speculations in the primary field of investigation'. Similarly, Toulmin (1953, pp. 38–39) regards a good model as experimentally fertile, suggesting further questions, taking us beyond the phenomena from which we began, and tempting us to formulate hypotheses. The 'intuitive grasp' (*Gestalt knowledge*) of the capacities and implications of a model is thus the key to the exploitation of its suggestive character.

Selectivity implies that models are different from reality in that they are *approximations* of it. A model must be simple enough for manipulation and understanding by its users, representative enough in the total range of the implications it may have, yet complex enough to represent accurately the system under study (Chorafas, 1965, p. 31). In another sense, too, models represent compromises in that each has a circumscribed range of conditions within which it has relevance (Skilling, 1964, p. 389A).

Because models are different from the real world they are *analogies*. The use of hardware models is an obvious example of the general aim of the model builder to reformulate some features of the real world into a more familiar, simplified, accessible, observable, easily-formulated or controllable form, from which conclusions can be deduced, which, in turn, can be reapplied to the real world (Chorley, 1964, pp. 127–128).

Reapplication is a prerequisite for models in the empirical sciences. Although some mathematical model builders disclaim responsibility for the degree to which their idealizations may represent the real world, claiming that their responsibility is discharged completely and with honour if they avoid internal error (Camp, 1961, p. 22); most geographical model builders would judge the value of a model almost entirely in terms of its reapplicability to the real world.

The functions of models. Models are necessary, therefore, to constitute a bridge between the observational and theoretical levels; and are concerned with simplification, reduction, concretization, experimentation, action, extension, globalization, theory formation and explanation (Apostel, 1961, p. 3). One of their main functions is *psychological* in enabling some group of phenomena to be visualized and comprehended which could otherwise not be because of its magnitude or complexity. Another is *acquisitive*, in that the model provides a framework wherein information may be defined, collected and ordered. Models have not only an *organizational* function with respect to data, but also a *fertility* in allowing the maximum amount of information to be squeezed out of the data (see the 'statistical models' of Krumbein and Graybill, 1965). Models also perform a *logical* function by helping to explain how a particular phenomenon comes about. The question as to what constitutes a satisfactory explanation is a complex one, but Bridgman (1936, p. 63) put it in model terms when he wrote; 'Explanation consists of analysing our complicated systems into simpler systems in such a way that we recognize in the complicated systems the interplay of elements already so familiar to us that we accept them as not needing explanation'. Models also perform a *normative* function by comparing some phenomenon with a more familiar one (Hutton, 1953–54, pp. 285–286). The *systematic* function of model building has already been stressed in which reality is viewed in terms of interlocking systems, such that one view of the history of science is that it represents the construction of a succession of models by which systems have been explored and tested (Meadows, 1957, p. 3). This leads to the *constructional* function of models in that they form stepping stones to the building of theories and laws. Models and theories are very closely linked (Theobald, 1964, p. 260), perhaps differing only in the degree of probability with which they can predict reality. The terms 'true' or 'false' cannot usefully be applied in the evaluation of models, however, and must be replaced by ones like 'appropriate', 'stimulating' or 'significant'. Laws are statements of very high probability and, as such, all laws are models, but not all models are laws. Finally there is the *cognative* function of models, promoting the communication of scientific ideas. This communication 'is not a matter merely of the sociology of science, but is intrinsic to its logic; as in art, the idea is nothing till it has found expression' (Kaplan, 1964, p. 269).

Types of models. Chorley (1964) provided an initial structure for the classification of models currently used in geography and this 'model of models' has been expanded and revised with special reference to geomorphology in Chapter 3 (Fig. 3.1).

The term 'model' has been used, however, in such a wide variety of contexts that it is difficult to define even the broad types of usage without ambiguity. One division is between the *descriptive* and the *normative*; the former concerned with some stylistic description of reality and the latter with what might be expected to occur under certain stated conditions. Descriptive models can be dominantly *static*, concentrating on equilibrium structural features, or *dynamic*, concentrating on processes and functions through time. Where the time element is particularly stressed *historical* models result. Descriptive models may be concerned with the organization of empirical information, and be termed *data, classificatory (taxonomic)*, or *experimental design* models (Suppes, 1962). Normative models often involve the use of a more familiar situation as a model for a less familiar one, either in a time (*historical*) or a *spatial* sense, and have a strongly *predictive* connotation.

Models can also be classed according to the stuff from which they are made, into, firstly, *hardware, physical* or *experimental* constructions, and, secondly, into *theoretical, symbolic, conceptual* or *mental* models. The former can either be *iconic* (Ackoff, Gupta and Minas, 1962), wherein the relevant properties of the real world are represented by the same properties with only a change in *scale*, or *analogue (simulation)* models, having real-world properties represented by different properties. The latter are concerned with symbolic or *formal* assertions of a *verbal* or *mathematical* kind in *logical* terms (Rosenblueth and Wiener, 1944–45, p. 317; Beament, 1960). Mathematical models can be further classed according to the degree of probability associated with their prediction into *deterministic* and *stochastic*.

Another view of models concentrates upon them as *systems* which can be defined on the basis of the relative interest of the model builder in the input/output variables, as distinct from the internal status variables. In order of decreasing interest in the status variables, many models can be viewed as *synthetic systems, partial systems* and *black boxes*.

The scale on which models are valuable and the standpoint from which they are constructed allow further distinctions, notably into *internalized* models which give a very parochial view of reality, and *paradigms* which are broadly significant models of value to a wide community of scholars.

Pitfalls in model building. The characteristics of models imply the existence of many dangers to which the model builder may fall prey. Simplification might lead to 'throwing the baby out with the bath water'; structuring to spurious correlation; suggestiveness to improper prediction; approximation to unreality; and analogy to unjustifiable leaps into different domains. Kaplan (1964, pp. 275–288) has summed up many of the dangers as problems of *overemphasis* on symbols, form, simplification, rigor and prediction. According to this view, a bad model would be heavily symbolic, present an overly-formalized view of reality, be much over-simplified, represent an attempt to erect a more exact structure than the data allows, and be used for inappropriate prediction.

Many philosophers have pointed to the dangers of craving for generality and of adopting a contemptuous attitude towards the particular case. They have often considered reality to be of too complex and multivariate a character to be susceptible to reasoning by analogy, and have asked whether the use of models introduces too great a detour into the reasoning process. In short, some hold that we should take heed of the Second Commandment: 'Thou shalt not make unto thee any graven image, or any likeness of anything that is in the heaven above, or that is in the earth beneath, or that is in the water under the earth'. In reply to this view of model building Ubbink (1961, p. 178) asks 'if this should be the case: in what sense can knowledge be true?'. Model building and reasoning are indissoluble, but 'the price of the employment of models is eternal vigilance' (Braithwaite, 1953, p. 93). Kaplan (1964, p. 276) believes that such vigilance is all the more necessary when model building is currently fashionable: 'The danger is all the greater with respect to model building because so much else in our culture conspires to make of it the glass of fashion and the mould of form. Models seem peculiarly appropriate to a brave new world of computers, automation and space technology, and to the astonishing status suddenly accorded to the scientist in government, industry and the military. It is easy to feel drawn to the wave of the future, and such tides are flowing strong today.'

The nature of paradigms

Paradigms may be regarded as stable patterns of scientific activity. They are in a sense large-scale models, but differ from models in the sense used above in that: (1) they are rarely so specifically formulated; and (2) they refer to patterns of searching the real world rather than to the real world itself. Scientists whose research is based on shared paradigms are committed to the same problems, rules and standards, i.e. they form a continuing community devoted to a particular research tradition. In a sense then, paradigms may be regarded here as 'super models' within which the smaller-scale models are set. As such, Thomas Kuhn in his *Structure of Scientific Revolutions* (1962), has assigned to the origin, continuance, and obsolescence of paradigms a prior place in the history of the evolution of science. Progress in research requires the continual discarding of outdated models, and subsequent remodelling. The more internally consistent the original paradigm or model, the more difficult it may be to remodel an existing structure in step with changing notions and increasing data. It is usual, therefore, for the most significant intellectual steps to be marked by the emergence of completely new models (Skilling, 1964, p. 389A). As Kuhn (1962, p. 17) has argued '. . . no natural history can be interpreted in the absence of at least some implicit body of intertwined theoretical and methodological belief that permits selection, evaluation, and criticism'. Without such paradigms all the available facts may seem equally-likely candidates for inclusion. As a direct consequence there is: (1) no case for the highly-defined, fact-gathering so typical of the exact sciences; and (2) a tendency to restrict fact-gathering to the wealth of available data which comes easily to hand. The fact that much of this data is a secondary by-product of administrative systems adds further to the massive data-handling problem. Certainly most geographical accounts are strongly 'circumstantial', juxtaposing facts of theoretical interest with others so unrelated or so complex as to be outside the bounds of available explanatory models.

The importance of the paradigm lies then, in Kuhn's terms, in providing rules that: (1) tell us what both the world – and our science – are like; and (2) allow us to concentrate on the esoteric problems that these rules together with existing knowledge define. Paradigms tend to be, by nature, highly restrictive. They focus attention upon a small range of problems, often enough somewhat esoteric problems, to allow the concentration of investigation on some part of the man-environment system in a detail and depth that might otherwise prove unlikely, if not inconceivable. This concentration appears to have been a necessary part of scientific advance, allowing the solution of puzzles outside the limits of pre-paradigm thinking.

In practice such 'rules' are acquired through one's education and subsequent exposure to the literature, rather than being formally taught. Indeed a concern about them only comes to the fore when there is a deep and recurrent insecurity about the nature of the existing paradigm. Methodological debates, concern over 'legitimate' problems or appropriate methods of analysis are symptomatic of the pre-paradigm period in the evolution of a science. Once the paradigm is fully established the debate languishes through lack of interest or lack of need. Thus in contemporary economics, the most successful and sophisticated of social sciences, the early debates over the nature of economics have been replaced by rather stable – but largely invisible – rules as to what problems and methods economic science should cultivate.

CLASSIFICATORY PARADIGMS IN GEOGRAPHY

Whatever the range of debate over the purpose and nature of geography, there is considerable communality of practice in the ways in which geographers have tackled their problems. Berry (1964) has analysed this paradigm of practice in terms of alternative approaches to a 'geographical data-matrix'. Here we look at his findings and attempt to diagnose the widespread unease generated by the continued use of this classificatory approach.

The geographical data matrix

Although regional geography, systematic geography, and historical geography are regarded as being quite distinct types of geographical study, Berry (1964) has deftly illustrated that each may be regarded merely as a different axis of approach to the same basic geographical data-matrix.

If a matrix has only *one* column, it is commonly called a 'column vector' (Krumbein and Graybill, 1965, p. 251), in which may be stored a series of j bits of information:

$$\begin{pmatrix} a_{11} \\ a_{12} \\ a_{13} \\ a_{14} \\ \cdot \\ a_{1j} \end{pmatrix}$$

Similarly we may store information about j elements (i.e. temperatures, elevations, population densities, etc.) in a regional column, to give an inventory of all the available characteristics of a given location.

A matrix with only *one* row is termed a 'row vector'. Here we may store a series of i bits of information:

$$(a_{11} \quad a_{21} \quad a_{31} \quad a_{41} \quad \cdot \quad a_{i1})$$

In this approach we store information about the same element but we vary the location to give the standard pattern of systematic geography, i.e. the mapping of a single feature (e.g. population densities).

By combining both the set of regions $(1 \ldots i)$ and the set of elements $(1 \ldots j)$ we have a rectangular array of the form:

$$\begin{pmatrix} a_{11} & a_{21} & a_{31} & a_{41} & \cdot & a_{i1} \\ a_{12} & a_{22} & a_{32} & a_{42} & \cdot & a_{i2} \\ a_{13} & a_{23} & a_{33} & a_{43} & \cdot & a_{i3} \\ a_{14} & a_{24} & a_{34} & a_{44} & \cdot & a_{i4} \\ \cdot & \cdot & \cdot & \cdot & \cdot & \cdot \\ a_{1j} & a_{2j} & a_{3j} & a_{4j} & \cdot & a_{ij} \end{pmatrix}$$

This matrix or box is termed by Berry the *geographical data-matrix* in that items containing information about the earth's surface may be stored in terms of their *regional* (or locational) characteristics and their *elemental* (or substantive) characteristics. Table 1.1A gives a formal example of this sort of matrix, and Grigg (Chap. 12, below) discusses its logical basis in terms of regional models.

TABLE I.I

Transformation of Vectors in Geographical Data Matrices*

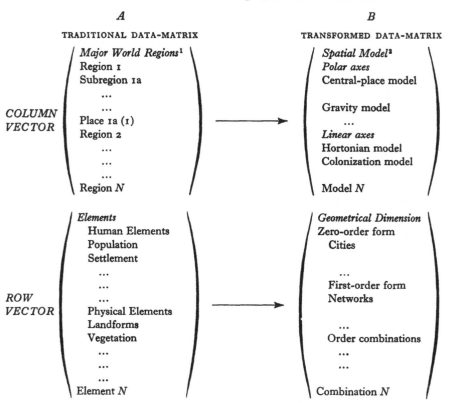

	A TRADITIONAL DATA-MATRIX	*B* TRANSFORMED DATA-MATRIX
COLUMN VECTOR	*Major World Regions*[1] Region 1 Subregion 1a ... ... Place 1a (1) Region 2 ... ... Region *N*	*Spatial Model*[2] *Polar axes* Central-place model Gravity model ... *Linear axes* Hortonian model Colonization model Model *N*
ROW VECTOR	*Elements* Human Elements Population Settlement ... ... Physical Elements Landforms Vegetation ... ... Element *N*	*Geometrical Dimension* Zero-order form Cities ... First-order form Networks ... Order combinations ... ... Combination *N*

By an ingenious series of row and column comparisons and by the addition of a third (time) dimension, Berry is able to reduce the great part of conventional geographical study to ten basic operations on the matrix. For example, areal differentiation is seen as column-vector comparisons and spatial covariance studies as row-vector comparisons. Comparison of a column over

---

* Adapted from Berry (1964, pp. 6–8).

[1] Information located with respect to *absolute* location (X, Y co-ordinates measured from a common base, e.g. latitude and longitude).

[2] Information plotted with respect to relative *location* (x, y co-ordinates measured from a variable base, e.g. distance and direction from a diffusion hearth).

time becomes sequent occupance, while other manipulations give the major modes of historical geography distinguished by Darby (1953). Indeed the major – and apparently fundamental – contrasts between regional, systematic and historical geography are seen by Berry (1964, p. 9) largely as a function of the relative length, breadth and depth of the study in terms of the three axes of the matrix.[1]

Difficulties of conventional matrix operations

The need for the information in the data matrix to be structured, given coherence, generalized and made intelligible has long been recognized, and as Wrigley (1965) points out, both Von Humboldt and Ritter scorned the attitude of their predecessors which had reduced geographical studies in the eighteenth century to mere 'pigeon-holing'. There is growing evidence however that neither of the two major solutions – the study of column vectors and of row vectors – effectively meet present-day demands.

(a) *The explosion of the data matrix.* The exact rate of growth of 'information' that could conceivably be stored in a geographical data-matrix is difficult to assess exactly. We have useful figures on the speed of mapping for selected areas (Langbein and Hoyt, 1959), and are familiar with the rapid accelerations caused by the development of air-photography after World War I and of satellite scanning and remote sensing since World War II. If we add to this the growing volume of statistical material collected by international agencies, national governments, state and local administrators, then the size of the potential world 'data bank' becomes staggering. If we restrict our data bank to one fact about each square-mile unit of the earth's surface then we have a basic store of 10^9 bits of information. If we relax this assumption to include the range of possible parameters for each unit (ranging, say, from a minimum of 1 to a maximum of 10^{10}) and a more finely-divided world grid (say 10^{11} units), then we may well have a storage problem of the order of 10^{11} to 10^{21} bits on our hands.

This is of course a speculative calculation, but what gives any such figure added point is the general rate of growth of information. Price (1963, pp. 4–13), after a wide-ranging survey of the growth of scientific information, found evidence of an exponential growth of 'impressive consistency and regularity' – that is, the more information that exists the faster it grows. Depend-

[1] In more recent, and as yet unpublished work, Berry has developed a second data-matrix (an interaction matrix) in which pairs of locations (*dyads*) occupy the rows and interactions occupy the columns. By reducing the two matrices through factor analysis, the general relationships between spatial structure (Matrix I) and spatial interaction or behaviour (Matrix II) may be explored. Through this extension Berry has not only clarified the relationships between formal and functional regions, but has also laid the basis for a much more general 'field-theory' of spatial behaviour.

ing upon what we measure, it is possible to estimate that the amount of information tends to double within a period of 10 to 15 years – probably slightly shorter. If one accepts the general form of the curve then the problem facing Humboldt and Ritter was about 1/1,000 times as small as that facing the current generation of geographers. The fact that large areas of the world were still then 'blank on the map' suggests this is in fact a sizeable overestimate; that is, the rate of growth of locationally 'storable' information has been *more* rapid than that of scientific information as a whole as the ecumene itself has expanded. Stoddart (1967) has suggested that, although the intensity of geographical data generation and handling over the past 200 years has been roughly half that of science as a whole (Fig. 1.2), there are indications of a current increase of activity.

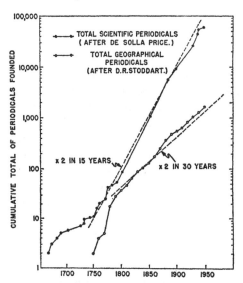

1.2 Cumulative totals of scientific and geographical periodicals founded. The dashed lines represent doubling in 15 and 30 years, respectively. (*Source: D. R. Stoddart. See also Stoddart, 1967*).

(b) *The weakening of conventional vector analysis.* Continuing Berry's (1964) matrix algebra analogy, we may trace the problems in columnar analyses to: (1) the expanding number of locational columns as old areas were ever more finely divided and new areas added; and (2) the proven failure of local 'column' factors to explain the juxtaposition of other variables in the same column. Wrigley (1965) has shown, for example, how the development of general production techniques during the Industrial Revolution tended to destroy unique regional features and to make the geographical characteristics of these regions less explainable in purely local terms. Analysis of row vectors (i.e. systematic distribution studies) has been weakened by the tendency for individual systematic sciences to 'split off' their own rows for separate analyses. It is significant, for example, that systematic sciences are increasingly taking on their own mapping analysis – and in some cases (e.g. Perring and Walters, 1962) are themselves making substantial contributions to cartographic technique.

Computer-orientated analysis of data matrices

One trend which runs counter to the information explosion is the increasing adaptation of digital computers to this problem. The whole field of data storage, data retrieval, data analysis, and data display is expanding rapidly and we may expect that geography, as a special type of *regional data-storage system*, will reap considerable benefits from this field of electronic technology.

Strides at the moment are being made in reducing the size and complexity of the data matrix through general data-evaluation and pattern-search programmes, some of which involve factor-analysis, cluster analysis and related multivariate techniques (Krumbein and Graybill, 1965, Chap. 15). WHIRLPOOL is a typical example of a powerful sequential multiple linear regression programme (Krumbein, Benson and Hempkins, 1964) for 'sorting out' significant factors in the series of elements in the matrix. An alternative approach to the regional columns of the data matrix is being led by Berry (1961; Steiner, 1965) in using a linked series of factor analysis, D^2 analysis, and discriminant analysis to 'optimize' regional divisions; thereby 'collapsing' the number of regional vectors into the smallest or most convenient number consistent with a given level of information (Haggett, 1965, p. 256). Other approaches have concentrated on the collapsing of the detailed isarithmic map into a series of *regional trends*. Chorley and Haggett (1965) have shown how orthogonal polynomials may be used to store vast amounts of regional information on the undulating isarithmic surface in terms of a very few basic equations, while Tobler (1966) has analysed other forms of numerical mapping and rapid map display.

It is clear then that technological improvements, largely in computer engineering and programming logic, are at last beginning to restore to the greatly-strained geographical data-matrix some of the order that its rapid expansion threatened to destroy. While empirical approaches to the stored material are likely to throw up a good deal of unexpected patterning, we should recall that these programmes are '. . . greatly facilitated when preceded by analysis of the problem in terms of some conceptual model – a mental picture – that defines the class of objects or events to be studied, the kinds of measurements to be made, and the properties or attributes of these measurements' (Krumbein and Graybill, 1965, p. viii). It is to these 'mental pictures' that we turn again in our last section.

TOWARDS A MODEL-BASED PARADIGM OF GEOGRAPHY

The continuing problems of the traditional class-based paradigm of geography suggest that experimentation with alternative approaches might prove fruitful. Here we: (1) attempt to isolate some of the components in a model-based paradigm; (2) discuss the impact it might have on research; and (3) inquire whether it meets certain basic requirements that a new 'candidate' should possess.

A model-based paradigm: a proposal

We do not propose to alter the basic Hartshorne (1959, p. 11) definition of geography's prime task, nor challenge the appropriateness of the matrix concept. We suggest, however, that it may be possible to derive a *second* matrix from the first by transforming the two basic vectors in such a way as to throw emphasis away from that of classification towards model-building. Table 1.1B gives an example of such a transformed matrix.

(a) *Vector I: locational relativity.* The search for greater accuracy along the absolute locational vector (e.g. exact latitude and longitude), and the search for significant 'fixes' in this absolute regional space (e.g. the location of the source of the Nile) is now largely centred in specialized government mapping agencies outside the university research world. However, location in an absolute sense on an isotropic surface is clearly only one way of viewing space. Bunge (1962) has urged that quite different and alternative ways are both possible and common, e.g., most ideas on *accessibility* or *isolation* refer to distance measured in a specialized way (usually in terms of *energy* translated into cost terms) and from specialized origins or axes. The selection of these *axes of symmetry* is largely determined by the spatial model being adopted: in a diffusion model the appropriate axis might be the diffusion *hearth*, and the appropriate parameter distance (accessibility) from that axis. Many of the models featured in later chapters use just such relativistic ideas of location with space measured in metric terms (see Henshall's discussion on Von Thünen's model in Chap. 11, below) or topological terms (see Haggett's review of Pitt's accessibility model in Chap. 15, below), with the measurements taken from single points, sets of points, or linear axes (e.g. Horton's model of erosion in terms of distance from the watershed). Attempts to use *relative* location rather than absolute location are a feature of spatial model-building. Although no formal attempts have yet been made to build up a 'general theory of locational relativity', the multivariate analysis of distance (measured in 'real' energy terms rather than 'neutral' *mileage* terms) and their reduction to suitable display maps pose fascinating and complex research problems. A whole new family of atlases might be envisaged showing completely different spatial patterns than the familiar absolute patterns of traditional maps. As Bunge (1964) has pointed out, on a cost-transformed surface pre-railway England reduces to coalfields grouped around an *inland* sea!

(b) *Vector II: topological-geometrical form.* The successive sub-division of the 'elements' vector has seen the constant departure of newly fledged sciences with their own geographical interpretations. Here we seem to have missed a fundamental point that these elements (e.g. vegetation, population) are not *in themselves* the objects of geography. Sten de Geer (1923, p. 2) argued that '... it is certain abstract qualities in the objects which are studied by Geography, and not the objects themselves'. We propose therefore that certain 'abstract properties', – the topological and geometrical form of that object or objects – be substituted for the standard 'element-class' vector. Cole (1966) has extended point-set theory to suggest that using a simple dimensional classification into zero-order (points, cities), first-order (lines, networks), second-order (areas, states), and third-order (surfaces, terrain) provides such an appropriate topological vector. Bunge (1962, p. 197) concludes his *Theoretical Geography* on a similar vein: 'the profession might, as a matter of efficiency, start dividing itself into various theoretical spatial fields, such as point problems, area problems, description of mathematical surfaces, and central place problems, rather than the current arrangement of climatology, population geography, landforms, etc.' Despite the apparent simplicity of this schema it can be rigorously extended (by set combinations) to include complex combinations of the basic topological forms: thus the four basic dimensions give fifteen combinations (Cole, 1966; see his Fig. 3.1). If we divide each of the four into only two time states (*stable* and *unstable*) then the number of possible combinations goes up to over 250. We have argued elsewhere that '.... many of the more successful attempts at geographical models have stemmed from this (geometrical) type of analysis' (Chorley and Haggett, 1965, p. 376) and will not repeat those arguments in detail. One such topological class – the linear network – is currently being investigated (see Haggett, Chap. 15, below; Haggett and Chorley, In preparation). In general we feel that geometrical analysis offers a logical, consistent and geographically more relevant alternative to the 'element-orientated' approach with its inevitable tendency to sub-divide geography and force it outwards towards the relevant external systematic disciplines. It not only offers a chance to weld human and physical geography into a new working partnership, but revives the central role of cartography (see Board, Chap. 16, below) in relation to the two.

Bernal (1965, p. 97) traces the change in biology through phases when it was 'primarily a descriptive science, more like geography, dealing with the structure and working of a number of peculiarly organized entities, at a particular moment of time on a particular planet', through its fission into zoology, botany and their subdivisions, and its present convergence towards common concern with chemistry and physics in molecular biology, cell biology, biophysics, biochemistry, etc. Geography has likewise shown a tendency to replace its first phase of fact gathering with a fission into physical and human geography (and their subdivisions). Whether there are signs within contemporary geography of a third converging stage are matters of debate. Certainly the general concept of systems analysis (Chorley, 1962) has been put forward by Ackerman (1963) as a fundamental integrating concept for geography, and Stoddart (1965; see also Chap. 13, below) has analysed its particular role in relation to ecological systems (ecosystems). Whatever the problems that remain to be solved, particularly problems of system identification and energy monitoring, we may expect that regional systems analysis will emerge as a major theme in geographical work over the next decade. It is not without interest that two of the most tractable systems identified and analysed by geographers – the watershed system (see Chorley, Chap. 3, below: and More, Chap. 4, below) and the nodal 'city region' (Garner, Chap. 9) – are distinctive geometrical forms with particular mathematical properties (e.g. nesting) and are organized with respect to specific axes from which relative distances are measured. That is, we may argue that distinctive *geosystems* (Stoddart, Chap. 13, below) are likely to be found at the intersections of distinctive vectors within our transformed data matrix.

It would be irrelevant here to pursue the applications of our transformation model to detailed cases. Table 1.2 (overleaf) is included to show the organization used by one of the writers in dealing with a typical nodal-system in human geography: it is based largely on topological distinctions (i.e. Vector I organization).

Implications for research

Kuhn (1962) has shown that the period which follows the wholesale adoption of a paradigm is commonly devoted to three main classes of research problem: (1) the determination of significant facts; (2) the matching of facts with theory; and (3) the further articulation of theory.

(1) One of the curious effects of paradigm adoption is to replace a period of data abundance with one of data scarcity. The very definition of the major research problem highlights the paucity of *relevant* data and great attention is paid to the most accurate measurement of constants that would have seemed, in the pre-paradigm period, to be either too recondite or too elusive to measure. Thus a study of urban-centred fields has led to the most careful study of city sizes using stringent operational criteria, and the need for observational networks of a new rigour and accuracy makes increasing demands on research time. With further concentration we might expect that problems

TABLE 1.2
Three-Stage Model for the Analysis of Regional Systems*

	STAGE I *System Identification*	STAGE II A. *Form Differentiation (Static)*		STAGE II B. *Form Differentiation (Dynamic)*		STAGE III *System Integration*
Dimensional Number	$\{0, 2\}$	0	1	3	4	$\int_0^4$
Geographical Form	City (*polar axis*) City region (*boundaries*)	Cities Settlements Urban hierarchies	Transport networks Communication systems	Urban fields Density gradients Land-use intensity	Innovation waves Frontier movements Sequent occupance Colonization	Regional systems Internal feedbacks Interregional systems External feedbacks
Analytical Techniques	Numerical taxonomy Local residuals Regional analogues	Rank-size analysis Nearest-neighbour analysis Quadrat analysis	Graph-theoretic analysis Connectivity Network geometry	Trend-surface analysis Harmonic analysis Fourier analysis	Physical simulation Monte Carlo models Markov-chain models	Matrix analysis, factor analysis Input-output analysis Interregional linear programming
Spatial Model	Regional hierarchies Formal, functional regions	Central-place theory Gravity models Weberian models Basic non-basic models	Network models Random graph models Geodesic models	Gravity models Absorption models Intervening-opportunity models Von Thünen models Potential models	Diffusion models Migration models Colonization models	Regional climax models Regional multipliers Growth poles
Major Sources for Spatial Models	Decision theory (Psychol.) Taxonomy (Biol.) Discriminant analysis (Stats.)	Point set theory (Math.) Organization models (Manag.) Packing theory (Math.)	Graph theory (Math.) Circuit design (Electr.) Search theory (Math., Psychol., Zool.)	Least-effort models (Sociol.) Minimum-energy models (Phys.) Potential models (Phys.) Game theory (Psychol.)	Epidemic theory (Medic.) Diffusion theory (Fluid dyn.) Rumour theory (Sociol.) Colonization & succession models (Bot.)	General Systems theory (Biol.) Ecosystems (Biol.) Interregional trade theory (Econ.) Multiplier models (Econ.)

* The table is schematic only: it follows the general arrangement of topics in Haggett (1965), Chaps. 2–6 inclusive.

of interregional flow and internal energy flux would put data collection and data standardization as major consumers of research time. In the case of physical sciences such problems are commonly accompanied by the design and construction of measuring apparatus of increasing precision and complexity. The analogy in non-experimental sciences may well be the development of computer programmes for squeezing the greatest amount of information from limited data.

(2) Determination of facts that can be directly related to paradigm models form a second class of activity (Kuhn, 1962, p. 33). Although such facts cannot be said to test the model in any strict sense, they provide a focus for research areas. Thus Gunawardena (1964) directed attention at the central-place structure of southern Ceylon in an attempt to test existing central-place models. Much of the stimulus that a model provides is the drive to test (and possibly overthrow) existing models. The strength of the paradigm is that it allows the progressive evolution of models within its general terms. Kuhn argues that only when no new models (or wholly incompatible new models) can be produced is the paradigm itself likely to shift.

(3) Extension and articulation of theory therefore provides the third class of research. This may take a number of forms but will probably include: (a) the determination of constants in existing predictive equations; (b) the quantification or further mathematization of existing qualitative models; and (c) the speculative and exploratory extension of extant models into areas of unproven and even unlikely application. Olsson's (1965) study of migration rates might fall into the first category, Dacey's (1965) elegant extension of Christaller-Lösch settlement models through point-set algebra into the second, and Gould's (1965) extension of search theory to the extension of transport networks in East Africa into the third.

According to the Kuhn thesis none of these major types of research is designed to produce results entirely outside the paradigm's limits, and the range of acceptable results from the studies is small – certainly small in relation to the results that could be conceived. The internal discipline of the paradigm, its unwritten rules and traditions, guides the pattern of research and ensures by the successive cumulation of small highly-limited advances that the science as a whole will progress.

Needed characteristics of a new paradigm

How should we recognize an efficient paradigm if we saw one? Kuhn (1962, pp. 152–158) has studied the introduction of new paradigms in fields as unlike as chemistry and electricity, and suggests three minimum ingredients for success.

(1) Firstly, the new paradigm must be able to solve at least some of the problems that have brought the old one to crisis point. We have argued above

that the most sorely-troubling feature of present geography is the explosion of the traditional data-matrix, and the forcing of geographers to study both areas and topics less and less relevant to the general shaping of the earth surface they profess to study. A new paradigm in geography must be able to rise above this flood-tide of information and push out confidently and rapidly into new data-territories. It must possess the scientific habit of seeking for relevant pattern and order in information, and the related ability to rapidly discard irrelevant information: '. . . it is the capacity for pattern-seeing and not the actual surveying of the landscape which explains this rapidity' (of scientific development). 'It explains why scientific activity remains a mystery for those devoid of theoretical insight, who see only facts' (Van Duijn, 1961, p. 67).

(2) Secondly, the new paradigm must appeal to the workers' sense of what is elegant, appropriate and simple. This somewhat aesthetic characteristic is difficult to define in specific terms and is most clearly seen in the mathematician's attraction towards elegant rather than inelegant mathematical proofs. At the smaller scale, the demonstration that Horton's law of stream numbers was a simple combinatorial system applicable to a much wider range of phenomena (Haggett: Chap. 15, below) represents an appropriate and economical simplification. Similarly Hägerstrand's (1953) overview of diffusion waves replaced many separate, clumsy, and individually articulated 'frontier' and 'sequent occupance' studies. A new paradigm for geography needs to provide a similar economical and elegant simplification for the *whole* field.

(3) Thirdly, the new paradigm must contain more 'potential for expansion' than the old. This characteristic is believed by Kuhn to be often the decisive one, albeit its adoption is based on faith in the new rather than its proven ability. Geography, coming late to the paradigm race, has the compensating advantage that it can study at leisure the 'take-off' paradigms of other sciences. There is good reason to think that those subjects which have modelled their forms on mathematics and physics – themselves 'leading sectors' in the scientific community (to continue Rostow's language) – have climbed considerably more rapidly than those which have attempted to build internal or idiographic structures. Not a little of this success stems from the great elasticity of mathematical analysis and the hierarchy of ever-more-complex equations that can be derived for observed patterns. As Kaplan (1964, p. 262) remarked: 'The use of mathematics and the construction of logical systems marks a certain coming of age'.

EPILOGUE

In this introductory chapter we have set out to examine the nature of models and their relation to facts on the one hand and to paradigms on the other. We have looked at the traditional paradigmatic model of geography and suggested that it is largely classificatory and that it is under severe stress. We have tentatively suggested an alternative model-based approach.

In judging the success of this approach and of the models discussed in the chapters which follow, we should recall that geography must measure its progress by the number of puzzles it has effectively solved, not by the magnitude of those that remain unsolved. In welcoming Ackerman's (1963, p. 435) reminder that the philosophical goal of geography is '. . . nothing less than an understanding of the vast, interacting system comprising all humanity and its natural environment on the surface of the earth' we should recall, with Humboldt, that such a goal is utterly unattainable in any complete sense – either now or in the future. Successful application of models in geography ensures no teleological progress towards full understanding, for scientific effort does not reduce the sum total of problems to be solved – it rather increases them.

REFERENCES

ACKERMAN, E. A., [1963], Where is a research frontier?; *Annals of the Association of American Geographers*, 53, 429–440.

ACKOFF, R. L., GUPTA, S. K. and MINAS, J. S., [1962], *Scientific Method: Optimizing Research Decisions*, (New York), 464 pp.

APOSTEL, L., [1961], Towards the formal study of models in the non-formal sciences; In Freudenthal, H., (Ed.), *The Concept and the Role of the Model in Mathematics and Natural and Social Sciences*, (Dordtrecht, Holland), 1–37.

BAMBROUGH, R., [1964], Principia Metaphysica; *Philosophy*, 39, 97–109.

BEAMENT, J. W. L., (Ed.), [1960], *Models and Analogues in Biology*; Symposium No. 14 of the Society for Experimental Biology, (Cambridge), 255 pp.

BERNAL, J. D., [1965], Molecular structure, biochemical function, and evolution; In Waterman, T. H. and Morowitz, H. J., (Eds.), *Theoretical and Mathematical Biology*, (New York), 96–135.

BERRY, B. J. L., [1961], A method for deriving multifactor uniform regions; *Przeglad Geograficzny*, 33, 263–282.

BERRY, B. J. L., [1964], Approaches to regional analysis: a synthesis; *Annals of the Association of American Geography*, 54, 2–11.

BLACK, M., [1962], *Models and Metaphors*, (Ithaca, New York), 267 pp.

BRAITHWAITE, R. B. [1953] *Scientific Explanation*, (Cambridge).

BRAITHWAITE, R. B., [1962], Models in the empirical sciences; In Nagel, E., Suppes, P. and Tarski, A., (Eds.), *Logic, Methodology and Philosophy of Science*, (Stanford), 224–231.

BRIDGMAN, P. W., [1936], *The Nature of Physical Theory*, (Princeton).

BROWN, L., [1965], Models for spatial diffusion research: a review; *Office of Naval Research, Geography Branch, Contract Nonr* 1288 (33), *Technical Report*, 3.

BUNGE, W., [1962], Theoretical geography; *Lund Studies in Geography, Series C, General and Mathematical Geography*, 1.

BUNGE, W., [1964], Geographical dialectics; *Professional Geographer*, 16 (4), 28–29.

CAMP, G. D., [1961], Models as approximations; In Banbury, J. and Maitland, J., (Eds.), *Proceedings of the Second International Conference on Operational Research*, (Aix-en-Provence), 20–25.

CAWS, P., [1965], *The Philosophy of Science*, (Princeton), 354 pp.

CHORAFAS, D. N., [1965], *Systems and Stimulation*, (New York), 503 pp.

CHORLEY, R. J., [1962], Geomorphology and general systems theory; *U.S. Geological Survey, Professional Paper 500-B*, 10 pp.

CHORLEY, R. J., [1964], Geography and analogue theory; *Annals of the Association of American Geographers*, 54, 127–137.

CHORLEY, R. J. and HAGGETT, P., [1965], Trend-surface mapping in geographical research; *Transactions of the Institute of British Geographers*, No. 37, 47–67.

COLE, J. P., [1966], Set theory and geography; *Nottingham University, Department of Geography, Bulletin of Quantitative Data*, 2.

DACEY, M. F., [1965], The geometry of central place theory; *Geografiska Annaler*, 47B, 111–124.

DARBY, H. C., [1953], On the relations of geography and history; *Transactions of the Institute of British Geographers*, No. 19, 1–13.

DEUTSCH, K. W., [1948–49], Some notes on research on the role of models in the natural and social sciences; *Synthèse*, 7, 506–533.

GEER, S. DE, [1923], On the definition, methods and classification of geography; *Geografiska Annaler*, 5, 1–37.

GOULD, P., [1965], *A bibliography of space-searching procedures for geographers*; Pennsylvania State University, Department of Geography (Mimeographed).

GUNAWARDENA, K. A., [1964], Service centres in southern Ceylon; *University of Cambridge, Ph.D. Thesis*.

HÄGERSTRAND, T., [1953], *Innovationsförloppet ur korologisk synpunkt*, (Lund).

HAGGETT, P., [1965], *Locational Analysis in Human Geography*, (London).

HAGGETT, P. and CHORLEY, R. J., [1965], Frontier movements and the geographical tradition; in Chorley, R. J. and Haggett, P., (Eds.), *Frontiers in Geographical Teaching*, (London), 358–378.

HAGGETT, P. and CHORLEY, R. J., (In preparation), *Network models in geography*, (London).

HANSON, N. R., [1958], *Patterns of Discovery*, (Cambridge), 241 pp.

HARTSHORNE, R., [1959], *Perspective on the Nature of Geography*, (London).

HUTTON, E. H., [1953–54], The role of models in physics; *British Journal of the Philosophy of Science*, 4, 284–301.

HESSE, M., [1953–54], Models in physics; *British Journal of the Philosophy of Science*, 4, 198–214.

KAPLAN, A., [1964], *The Conduct of Inquiry*, (San Francisco), 428 pp.

KRUMBEIN, W. C. and GRAYBILL, F. A., [1965], *An Introduction to Statistical Models in Geology*, (New York).

KRUMBEIN, W. C., BENSON, B. and HEMPKINS, W. B., [1964], WHIRLPOOL: a computer programme for 'sorting out' independent variables by sequential multiple linear regression; *Office of Naval Research, Geography Branch, Technical Report* 14, Task No. 389-135.

KUHN, T. S., [1962], *The Structure of Scientific Revolutions*, (Chicago).

KUIPERS, A., [1961], Model and insight; in Freudenthal, H., (Ed.), *The Concept and the Role of the Model in Mathematics and Natural and Social Sciences*, (Dordrecht, Holland), 125-132.

LANGBEIN, W. B. and HOYT, W. G., [1959], *Water Facts for the Nation's Future*, (New York), 228 pp.

LEWONTIN, R. C., [1963], Models, mathematics and metaphors; *Synthèse*, 15, 222-244.

MANLEY, G., [1966], A new geography; *The Guardian*, March 17th, 1966.

MEADOWS, P., [1957], Models, system and science; *American Sociological Review*, 22, 3-9.

NAGEL, E., [1961], *The Structure of Science*, (London), 618 pp.

OLSSON, G., [1965], Distance and human interaction: a bibliography and review; *Regional Science Research Institute, Bibliography Series*, 2.

PERRING, F. H. and WALTERS, S. M., [1962], *Atlas of the British Flora*, (London).

PORTER, P. B., [1954], Another puzzle-picture; *American Journal of Psychology*, 67, 550-551.

PRICE, D. J. DE SOLLA, [1963], *Little Science, Big Science*, (New York).

ROSENBLUETH, A. and WIENER, N., [1944-45], The role of models in science; *Philosophy of Science*, 11-12, 316-321.

SKILLING, H., [1964], An operational view; *American Scientist*, 52, 388A-396A.

STEINER, D., [1965], A multivariate statistical approach to climatic regionalization and classification; *Tijdschrift van het Koninklijk Nederlandsch Aardrijkskundig Genootschap*, 82, 329-347.

STODDART, D. R., [1965], Geography and the ecological approach: the ecosystem as a geographic principle and method; *Geography*, 50, 242-251.

STODDART, D. R., [1967], Growth and structure of geography; *Transactions of the Institute of British Geographers*, 41.

SUPPES, P., [1961], A comparison of the meaning and uses of models in mathematical and empirical sciences; In Freudenthal, H., (Ed.), *The Concept and the Role of the Model in Mathematics and Natural and Social Sciences*, (Dordrecht, Holland), 163-177.

SUPPES, P., [1962], Models of data; In Nagel, E., Suppes, P. and Tarski, A., (Eds.), *Logic, Methodology and Philosophy of Science*, (Stanford), 252-261.

THEOBALD, D. W., [1964], Models and method; *Philosophy*, 39, 260-267.

TOBLER, W., [1966]. Numerical map generalization; *Michigan Inter-University Community of Mathematical Geographers, Discussion Papers*, 8.

TOULMIN, S., [1953], *The Philosophy of Science*, (London).

UBBINK, J. B., [1961], Model, description and knowledge; In Freudenthal, H., (Ed.), *The Concept and the Role of the Model in Mathematics and Natural and Social Sciences*, (Dordrecht, Holland), 178-194.

VAN DUIJN, P., [1961], A model for theory finding in science; *Synthèse*, 13, 61-67.

WRIGLEY, E. A., [1965], Changes in the philosophy of geography; In Chorley, R. J. and Haggett, P., (Eds.), *Frontiers in Geographical Teaching*, (London), 3-20.

CHAPTER SIX

Demographic Models and Geography

E. A. WRIGLEY

INTRODUCTION

Population has always figured prominently in geography. Consideration of the distribution and density of population has been a starting point in many geographical studies, a finishing point in others, and has occasionally played both roles when the purpose of the study was to bring a new understanding of some aspects of distribution patterns (Wrigley, 1965B). We are frequently referred to maps showing densities of population, to general models of population distribution in space in the manner first pioneered by Christaller and Lösch, to studies of the relative rates of growth of settlements of different sizes, to questions like rural depopulation, to ideas like the market orientation of industry (which usually means simply the location of plants at places where there are a lot of people to buy the products). But for these purposes often little more than the crude totals of population are used, though sometimes these may be broken down further by occupational divisions, by age-structure, or in some other way. Changes in population totals over time and migration have occasionally been analysed with some rigour, and the use of models is now widespread in the attempt to achieve a clearer understanding of migration (see, for example, Lund Studies, 1957). Little attention, however, has been given to the construction or use of demographic models: nor has their potential importance to the understanding of several issues which bulk large in human geography been widely discussed.

Demographic models *sensu stricto* deal with the interplay of fertility, mortality and nuptiality. If these can be measured accurately demographic models will determine what the stable age-structure of the population will ultimately be assuming existing characteristics are maintained, what its rate of growth or decline, and so on. Alternatively, if they cannot be measured accurately, it may be possible with imperfect data to establish the extreme limits within which the truth must lie. Each variable will interact with the other two and produce sympathetic changes in them in the manner characteristic of open systems, and the changes can be predicted accurately if sufficient information is available. Much work of fundamental importance was carried out between the wars by Lotka in this field (Lotka, 1934–1939).

In a looser sense demographic models may be taken to include models in which the parameters may extend to features of the economic life of the community in question (United Nations, 1953; Leibenstein, 1954) or to its social and political activity (Sauvy, 1952–54). For example, to take a single restricted issue, use may be made of demographic models to determine the maximum percentage of matrilateral cross-cousin marriages which can occur in populations with a range of different demographic characteristics (Levy and Westoff, 1965). Or, more generally, a relationship between a system of land tenure and marriage customs may be postulated and the economic and demographic consequences of such a web of relationships can be examined by constructing a model in which the relationships between the major variables are expressed.

It is no accident that until recently models of these types aroused little interest in geographers, and for two reasons. In the first place it was commonly assumed that whatever the theoretical range of possibilities of demographic behaviour, most pre-industrial societies behaved in a manner which made it possible to treat population as a given feature of the situation whose demographic characteristics were of minor interest. This was the product of what might be called a modified Malthusianism. Few later writers have been so firmly pessimistic as the early Malthus in their treatment of the press of population against resources, but most have nevertheless tended to assume that pre-industrial populations normally grow towards a ceiling imposed by the maximum flow of food and raw materials which is possible at any given level of material technology and usable resources. In these circumstances it is clearly more important to devote attention to the nature of the material technology and the resource base of the community rather than to its demography. Another common assumption of a more general nature which also tends to cause a neglect of demographic issues is that which sees the main demographic characteristics of a community as a function of its fundamental economic constitution. Once more this implies that there is little point in investigating the structure of fertility, mortality and nuptiality in the population. If the general shape of the economic environment is known the chief features of the demographic situation can be read off from it.

Secondly, even when the treatment of population as a dependent variable was viewed with suspicion there was until recently a dearth of well-documented case studies to illustrate effectively the complexity of the interplay between demographic, economic, sociological and geographical variables. There is now a comparative abundance of these studies (e.g. Goubert, 1960; Hatt, 1952; Lorimer, 1954; Glass and Eversley, 1965). In one recent work, indeed, a most interesting attempt has been made to invert completely the Malthusian view that population size is dependent upon the potential of a given agricultural technology. It has been argued instead that agricultural technology is normally a function of population density. The dependent and independent variables have exchanged places (Boserup, 1965).

The importance and independence of the demography of a population has been brought home with special force to a wide public in recent years by the population problems of the developing countries. That they illustrate the point very dramatically, however, should not be allowed to obscure the fact that the case is general, not special.

ANIMAL POPULATION BEHAVIOUR

It may prove helpful to preface the discussion of demographic models by remarking the great change in thought about the characteristics of animal populations which has occurred recently. It forms a fascinating chapter in intellectual history and makes a convenient starting point for the subsequent discussion of the behaviour of men. The two types of population study have, after all, been closely linked in the past and have profited much from each other.

The study of animal population characteristics has long held a prominent place in biological studies. Darwin, acknowledging most generously the extent of his debt to Malthus, made of this an engine to drive the mechanism of natural selection. He pointed out that the long-term trend in the numbers of most animal populations was neither up nor down whereas their powers of reproduction were in all cases great enough to secure a rapid increase in their numbers. Even the elephant, according to Darwin's calculations, in spite of a very long gestation period, was capable of increasing from one couple to 15,000,000 in a period of 500 years (Darwin, 1869). Most animals possessed much more spectacular powers of multiplication.[1] The difference between the inherent powers of reproduction possessed by the animal kingdom and the absence of long-term increase represented in Darwin's view the pressure of selection, the constant sifting out and elimination of those individuals whose constitution made them less well able than their fellows to survive and reproduce. Those which were better adapted were able to breed and so the next generation included a higher proportion of individuals with the advantageous traits and the less well adapted gradually died out.

It is natural in terms of a model of this type to expect populations to be at

---

[1] '—reflect on the enormous multiplying power *inherent and annually in action* in all animals; reflect on the countless seed scattered by a hundred ingenious contrivances, year after year, over the whole face of the land; and yet we have every reason to suppose that the average percentage of every one of the inhabitants of a country will *ordinarily remain constant*.' (Darwin, 1958, p. 118.)

or close to the maximum which their habitat and their relative success in competition with other species permits them to attain. In recent years, however, it has come to be realized that this attitude needs substantial qualification if it is to do justice to the patterns of animal behaviour which are observable. The classical mechanism of selection may always be at work but is mediated through elaborate social conventions, many of which have developed in order to relieve a group of animals of one species living in an established territory from the extreme rigours of selection implied by the full use of the fertility potential of the individuals present. This view involves not only the assertion that mortality levels are dependent upon the density of the population – a point always recognized; but also that fertility levels vary with population density, falling as numbers rise and vice versa. In some forms of life the fall in fertility may be a simple reaction to direct physical problems (for example, the absence of further suitable sites for oviposition – there is a general discussion of this and other related points in Watt, 1962). In others the relationship is indirect and involves elaborate social conventions. In the former case the restriction of fertility may be viewed as an extension of the classical view of density-dependent mortality levels. In the latter, however, the social conventions, though closely geared to population densities, are very flexible and introduce a new element into the analysis of animal population behaviour. Populations of this type may fluctuate round a figure far below the maximum.

The conventions may take many forms. For example in birds which nest in colonies, such as storks or rooks, the number of nesting sites in the colony may be restricted by convention. Only those birds can breed which establish tenure upon one of the nesting sites. To secure a good site is evidence of high rank in the hierarchical society. Those lowest in the hierarchy, though sexually mature, may be unable to gain a site and so be prevented from breeding. The number of breeding pairs thus remains constant from year to year and the pressure upon the food supplies of the territory belonging to members of a rookery does not rise as it might if there were no social controls upon breeding. The mechanism of control varies in different species. For example, in the tits, each male seeks to establish control over an area from which he can exclude his rivals and which enables him to secure a mate, nest, and feed his young (see Wynne-Edwards, 1962, esp. Chap. 9). In other bird and animal species the social conventions governing fertility may take still other forms, but they all tend to the same end – to prevent numbers rising to the point where the means of subsistence is endangered by excessive population pressure or where adult members of the population are unable to keep themselves in good health because of the intensity of the competition for food. Selection operates, in short, at a social as well as an individual level, and the success of the group may be enhanced by the development of social habits which restrict fertility severely, or (which amounts to much the same thing) cause an extremely high mortality among the newly-born (see, for example, Wynne-Edwards, 1962, p. 537).[1]

Not all animal societies have developed mechanisms of this sort. Indeed the process of selection has thrown up radically different patterns of population behaviour in some cases, especially where the species in question lives in a marginal and difficult environment. For example, the locust attempts to maintain and extend its range by building up numbers very rapidly from time to time and 'exploding' into neighbouring areas. During the peak of a population cycle of this sort the food base available to the locust is exhausted over large areas, a pattern of events very different from that described in the previous paragraph. It has been supposed that the logic of population behaviour of this type may lie in the greater danger that exceptionally severe seasons in marginal areas will wipe out the whole of a local population so that waves of 'peak' populations spreading out in 'explosion' periods may find suitable areas to re-colonize. In more stable environments this is much less likely to occur and the long-term advantages of population 'explosions' is correspondingly less (see Wynne-Edwards, 1962, Chap. 20). At all events it is clear that a wide variety of types of population behaviour can be observed in animal populations, that a great deal of the social life of animals is connected with the fine adjustments of numbers to opportunities, and that it is far from being universally the case that the full reproductive potential of an animal group is called into play in normal seasons.

A SIMPLE DEMOGRAPHIC MODEL

What is true of animal societies is also true of societies of men and women. Societies have always, or almost always, developed social conventions which inhibited fertility to some extent, and often very drastically. Moreover, customs which have the effect of limiting fertility may be supplemented by others which cause high mortality among the very young. These have much the same effect as checks upon fertility in limiting the reproductive effort of the society. Levels of fertility and mortality, and particularly the latter, may

[1] Darwin was puzzled by the question of the age incidence of the heavy mortality which must follow from unrestricted fertility, but inclined to the view that it must fall mainly on the very young. For example, 'Where man has introduced plants and animals into a new country favourable to them, there are many accounts in how surprisingly few years the whole country was fully stocked; and yet we have every reason to believe from what is known of wild animals that *all* would pair in the spring. In the majority of cases it is most difficult to imagine where the check falls, generally no doubt on the seeds, eggs, and young; but when we remember how impossible even in mankind (so much better known than any other animal) it is to infer from repeated casual observations what the average of life is ... we ought to feel no legitimate surprise at not seeing where the check falls in animals and plants'. (Darwin, 1958, pp. 117–118).

fluctuate greatly from year to year and produce marked changes in a population in the short term, but the long-term effect of any prevailing level of fertility and mortality may be illustrated quite simply.

It is convenient to make a number of simplifying assumptions initially. Let us suppose that the society in question is at a comparatively primitive level of material culture, and that the material culture is static so that it is not possible to secure a steady increase in the production of food in the community by taking advantage of technical advances in agriculture. In these circumstances a population cannot grow through the ceiling set by the food-producing potential of the area at the given level of material culture. It is mistaken, however, to suppose that population will necessarily rise to this maximum. An equilibrium level of population may establish itself at many different

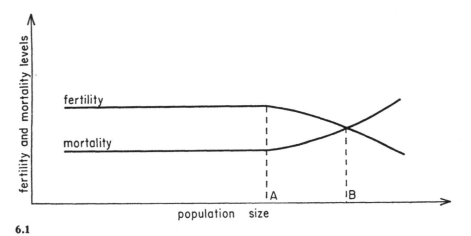

6.1

points. The level cannot, of course, be higher than the ceiling (except temporarily) and may well fall short of it by a substantial margin.

The combination of possibilities represented by the fertility and mortality schedules in Figure 6.1 describes a situation in which until population reaches A in total size the levels of fertility and mortality do not change and the population rises at a uniform speed. When population rises above A, however, both schedules are affected. Mortality rises and fertility falls as the density of population increases further until at B fertility and mortality are in balance and the population neither rises nor falls. There is, however, no a priori reason why fertility and mortality should both begin to be affected at the same density of population. In Figure 6.2 the pair of curves represents a situation closer perhaps to what Malthus had in mind in those moods when he doubted the efficacy of prudential checks (and which Darwin appears to have supposed typical of most animal populations). Here mortality follows the same path as in Figure 6.1, rising steadily when a certain critical density

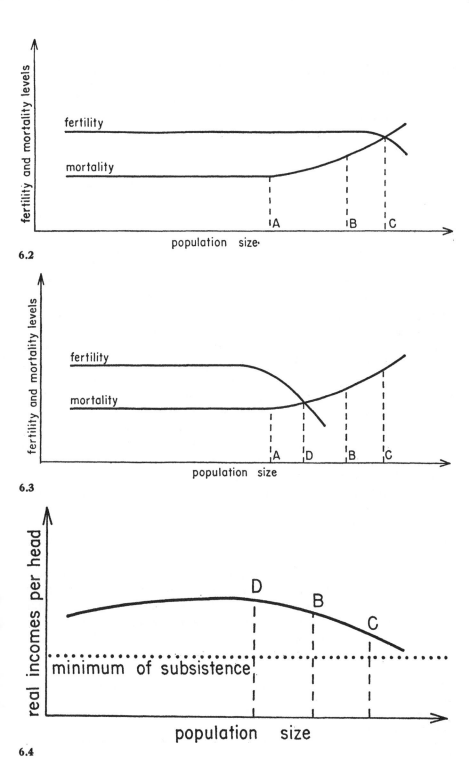

6.2

6.3

6.4

has been exceeded. But fertility is affected much later and less (indeed for the purpose of exposition, though not in reality, the fertility schedule might be a horizontal straight line). In these circumstances population will rise to C, a much higher level than B in Figure 6.1. Misery and vice, to use the language of Malthus, will be rife; or, in Darwinian terminology, the pressure of competitive selection will be very intense.

A third case may be imagined in which the mortality schedule is once more the same, but now fertility reacts very sensitively to any population pressure. The curve is deformed early and severely. This is shown in Figure 6.3. As long as this schedule describes the fertility characteristics of the population its total will never exceed D.

Figure 6.4 shows the implications of these three pairs of schedules in terms of real incomes per head. At any given level of material culture and with an inflexible resource base there may be expected to be some total of population at which real incomes per head will reach their maximum (see Sauvy, 1952–54 for a full discussion of this complex concept). It is reasonable to suppose that at some point after this optimum level has been passed mortality levels will begin to rise. If the fertility schedule intersects with the mortality schedule before the latter has begun to rise an optimum level of population may result enjoying the highest standard of living possible in the cultural and environmental context of the day and place. D, derived from Figure 6.3, represents a more advantageous position in Figure 6.4 than B which is the equilibrium level of total population from Figure 6.1, while C is the least happy position, reflecting the persistent high fertility found in Figure 6.2.

This account is both compressed and oversimplified. For example, it is obvious that mortality schedules may also vary considerably. In some societies infanticide was a common practice (Darwin thought Malthus had seriously underestimated its importance generally – Darwin, 1901, p. 69). It is conceivable that in such societies the mortality schedule might turn up as sharply as the fertility schedule turns down in Figure 6.3, with similar results as far as the equilibrium total of population is concerned.[1] Other qualifications and enlargements of the model will be made later in this essay, and many more would be called for in a larger treatment of the subject. But for the present purpose a sketch of this sort is sufficient if it makes it possible to discuss the importance to geography of the demographic chracteristics of populations. The elaboration of the model may now, therefore, be postponed in order to consider some of its implications in its present crude state.

---

[1] To calculate mortality from the moment of birth is, of course, a somewhat arbitrary proceeding. In all societies many lives are lost in the womb by miscarriage, abortion and stillbirth. In those societies in which infanticide is widely practised, it might be more useful for some purposes to measure mortality only from, say, the end of the first month of life, and to count death from infanticide, like stillbirth or abortion, as a feature which keeps fertility low rather than one which boosts mortality.

DEMOGRAPHIC CHARACTERISTICS AND GEOGRAPHICAL CONDITIONS

Many of the features of the human geography of an area which figure most prominently in orthodox discussions are in part a function of the demographic characteristics of the society in question. For example, it is evident that the subdivision of holdings may be carried much further in a given area when the population total is at, say, *C* in Figure 6.4 than would be the case if the total were at *D*. The pressure of high fertility cannot drive population increase up beyond a certain level because mortality will rise sufficiently to cause the increase to taper off, but it can push population totals up to a higher level than would be achieved at a lower level of fertility. This in turn may mean smaller holdings, more intensive use of the land (but at a lower output per worker), and the bringing into cultivation of land of poor quality (see Boserup, 1965, p. 118, for a lively exposition of the view that the ultimate result of an increasing density of population on the land may even be to better the prospects of sustained economic growth). Further examination of land use in the area may reveal other related effects. Where population pressure is great there may be a heavy premium placed on growing the crop which yields the largest return of calories to the acre – perhaps the potato; and the balance between livestock and crops may also be affected. Where it is necessary to maximize yield of food per acre, and where real incomes are low because the equilibrium level of total population is close to the Malthusian ceiling, livestock normally give way to crops and land that might otherwise produce good beef may be used instead for wheat (for a discussion of a very similar issue see Wrigley, 1962).

Through the same set of interrelated circumstances the local demography may have a great bearing on such matters as soil exhaustion, erosion, the upsetting of the hydrological balance, and so on. The progressive impoverishment of the soil leading ultimately to its destruction or to a permanent impairment of its fertility is not something which is solely or even primarily determined by the physical and chemical characteristics of the soil itself. It might indeed be said, if one wished to be paradoxical, that there is no marginal land *in se*. The manner in which the land is used is of paramount importance, and this in turn is in part a function of the pressure of population. Even the most unstable of soils and most delicately balanced ecological systems can be preserved if they are not subjected to undue pressure. Equally, even soils in areas not usually considered marginal may deteriorate if used without discretion. In late thirteenth-century England there are grounds for believing that population pressure at the contemporary level of agricultural technique was intense. There is indirect evidence that unsuitable land, lying on steep slopes or consisting of poor sands or chalk, was taken into cultivation for a short period but that the deterioration of the soil after a short period of use was so great that it dropped out of use and that as a result population fell. The Malthusian ceiling was probably closely approached with correspondingly severe effects (Postan and Titow, 1959: see also Herlihy, 1965, on the still more severe situation in northern Italy in the thirteenth century).

Or again it is easy to imagine an illuminating contrast in land use in a tropical area of slash-and-burn cultivation. In such areas it is impossible to keep the same piece of land in cultivation for more than a limited span of years because yields fall off sharply and the soils deteriorate fast under the strain of cultivation. Each small group of cultivators must therefore move on from time to time, and may need, say, twenty times as much land at its disposal as is in cultivation at any one time. Suppose that there are two groups possessing an identical material culture and living in the same environmental circumstances. They possess identical mortality schedules but different fertility schedules. It is quite possible to imagine that in one case the push of fertility may be strong enough to raise population to the point at which the group is cultivating at any one time not a twentieth of the total land in their possession but a fifteenth in order to meet their current food requirements. This will mean that each plot has insufficient time in which to recover from the last bout of cultivation. The land will yield progressively less and ultimately the population will be obliged to contract very severely and the land may be damaged permanently. The eroded land and exhausted soils which then occur in the area will have been damaged not for any reason intrinsic to their physical nature but because of the demographic characteristics of the group living there. The other group, with its lower fertility, may be able to continue to cultivate indefinitely and without danger land which was originally identical in nature.

FERTILITY LEVELS IN
PRE-INDUSTRIAL SOCIETIES

The assumption that the mortality schedule is necessarily the same in societies living at similar levels of material culture and comparable physical environments is, of course, unrealistic (even if it were justified it would still leave room for many different levels of mortality depending upon the degree of population pressure). But at any time before the Industrial Revolution it is reasonable to assume that mortality was less under social control than fertility, except in the case of the new-born in societies practising infanticide.

It may therefore be helpful at this point to review the range of social customs which may serve to keep fertility below the full potential physiologically available.

In western European societies during the centuries immediately before the Industrial Revolution the most important institutional check upon fertility was probably the conventional age at first marriage for women. This was normally in the middle twenties, and might occasionally be substantially higher. In the parish of Colyton in Devon, for example in the later seventeenth century it was about 30 (see Wrigley, 1966). Similar results have been obtained for the parish of Sainghin-en-Mélantois in northern France at the same period (see Deniel and Henry, 1965). Perhaps the lowest age that is well attested by family reconstitution methods is 22 during the period 1700–79 in a part of western Flanders (see Deprez, 1965A, pp. 615–616). Lower ages have been found occasionally but for special groups within the population rather than for a whole community (19.7 years for the British peerage in 1575–99, for example: Hollingsworth, 1964, p. 25). Since fecundity declines rapidly in the later 30s and last children are seldom borne at a mean age much above 40, it is obvious that a late age at first marriage may drastically reduce the number of children a woman bears. An average age at first marriage of 25, other things being equal, might well reduce total fertility by a third when compared with an average age at first marriage of, say, 18 – by no means uncommon in other places and periods. The effect of a late age at first marriage for women, moreover, is frequently compounded by a comparatively high percentage of women never marrying. These two demographic characteristics appear to be in most cases positively correlated with each other (see Hajnal, 1965).

The age pattern for women at first marriage found in pre-industrial western Europe, however, is seldom if ever found in other pre-industrial societies. In most African and Asiatic pre-industrial societies, for example, a high proportion of all women may be married before they attain the age of 20 and very few women remain spinsters for life (in India in 1891 the mean age at marriage of women was 12.5 years; by the age of 15 the vast majority of girls were married or widowed; see Goode, 1963, pp. 232–235). But there may be many other customs practised which serve to reduce fertility considerably. For example, in a society where adult mortality is high a ban upon the remarriage of widows will operate as a brake upon fertility. The custom of suckling children for two, three or four years after birth (even higher ages have been recorded quite frequently) may reduce marital fertility considerably, especially if combined with a taboo upon intercourse for a period after the birth of a child. It is possible that polygyny may have a similar effect. There is some evidence that women who are partners to a polygynous marriage may normally have a lower fertility than those monogamously married (see, for example, Dorjahn, 1958).

Many other methods of escaping the burdens of a too great fertility may also be found in pre-industrial societies. The extent and efficacy of methods of controlling conception before the nineteenth century are still a matter for debate and their impact may have been relatively slight in most societies (but see Wrigley, 1966, Carr-Saunders, 1922, and Lorimer, 1954). Abortion and infanticide, however, were common in many societies in the past. The latter is especially effective, from the point of view of limiting population growth, if the majority of the infants who are exposed or killed in other ways are female, and this was sometimes the case (Westermarck, 1921, vol. 3, pp. 58, 162 and 166–169). Infanticide may take many forms from the almost casual over-laying of young babies to the killing of those unfortunate enough to be born with physical characteristics which the society deemed abhorrent. In some societies those who were born with teeth already cut or who cut teeth in the upper part of the jaw before those in the lower were put to death: in others all twins (in general roughly one child in every forty is a twin), or those born feet first.

Sometimes the functional necessity for measures of this sort seems clear. For example, amongst some Australian aboriginal tribes, for whom ease of movement was essential, it was courting disaster for any mother to be burdened with more than one child not capable of keeping up with the tribe without assistance (see Spencer and Gillen, 1927, vol. I, pp. 39–40, 221; Krzywicki, pp. 119–144). The too-fertile woman was a figure of fun and if a woman became pregnant again too quickly the new child might well be destroyed at birth or very soon after. This was not a symptom of exceptional lack of feeling on the part of the aborigines. They treated the children whom they dared to allow to live with as much affection as children are treated in other lands and times, but long experience threw up this as a satisfactory solution to a problem which would otherwise have led to constant difficulty and perhaps danger for the group as a whole. In other cases there is no clear functional relationship between the immediate problems facing a society and practices which tend to lower effective fertility, either in the minds of members of those societies or visible to the anthropologist observing them. Nevertheless, social habits of this sort may well have become established, and have proved their selectional worth, because of their effect in making it easier to maintain population well below the maximum possible, just as many of the social habits of animals may prove to be best explained in this way.

A MODIFIED DEMOGRAPHIC MODEL

The model of the inter-relationship between demography and matters of prime interest to geography developed so far is too simple to be of much value. It must now be made to resemble reality a little more closely by modifying it in some particulars. The most important difference between the population characteristics of men and animals can be put very simply. It is legitimate to treat the long-term trend of any animal population as in principle horizontal. If environmental and ecological circumstances do not change, however much the totals of a species may fluctuate from one year to the next, it is reasonable to treat its mean number as unchanging. With men this is not so. Even before the Industrial Revolution very important changes in the material culture of societies took place and radically altered population maxima and optima from time to time. Where men are pegged down at a particular level of material culture for very long periods it is no coincidence that their demographic characteristics show a great similarity to those of animals. Most of the examples quoted in recent paragraphs to illustrate methods of population control were drawn from societies of rather low and unchanging material culture. The tribes of Australian aborigines met and came to terms with these problems in a manner which can be assimilated to general models of animal population behaviour, but matters become more complex once changes in material culture become sufficiently frequent and important.

Suppose, by way of illustration, that an agricultural group is settled in an area of mixed soils and relief in which chalk uplands alternate with clay vales. For many centuries they have been able to make effective use of the light chalk soils for agricultural purposes, but have entered the wooded clay vales only for hunting purposes. Over a period of centuries their population has tended to fluctuate round a given level, the equilibrium level of the fertility and mortality schedules of that society. Then a method of clearing the forest from the clay vales and cultivating the soil in suitable areas is developed or acquired from contact with a more advanced material culture. The capacity of the land to support people is raised considerably by this change, say, to three times its former level. New land for settlement is readily available.

It is possible to imagine population numbers rising to meet the new opportunities made available in this way without any change in the general level and shape of the fertility and mortality schedules, which would simply be displaced to the right, as in Figure 6.5. This would happen because food supplies and other necessities were now more abundant per head at levels of total population which would previously have meant severe hardship and rising mortality. For example, at P before the change in material culture mortality is rising fast. After the change a population of this size creates no difficulties (similar observations hold good for fertility also, of course). Population would continue to rise decade by decade until the new opportunities had been exhausted. Then mortality and fertility would again reflect population pressure and a new equilibrium level of population would be established at Y, replacing the old equilibrium level at X. The general situation, however, would be similar in all respects to that which had obtained previously except that the absolute number of individuals in the population had tripled.

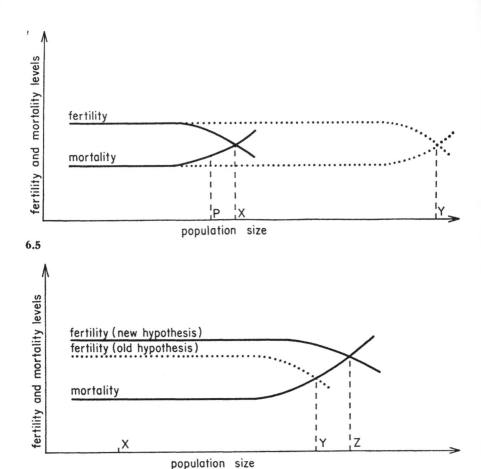

6.5

6.6

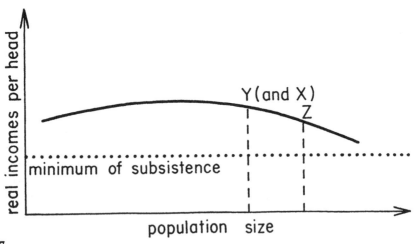

6.7

But in the new circumstances created by the change in material culture, things might fall out very differently. It might be, for example, that there would be changes in the level and shape of the fertility schedule in response to the new opportunities. These might occur because age at marriage for both sexes fell as a result of the abundance of good new land; or by the abandonment of practices tending to reduce fertility (perhaps a relaxation of attitude to the remarriage of widows; perhaps the abandonment of the procuring of abortions). Figure 6.6 shows a change of this sort. During the generations in which the clay vales were filling up, the new fertility schedule would carry with it no penalty in the form of lowered living standards such as would have been entailed if the change in fertility had occurred without any new economic opportunity. Nor would the soil itself be endangered by too fierce exploitation at this stage. But if, when the new population total implied by the new fertility schedule were approached at the end of the period of colonizing the clays, the new fertility characteristics had become firmly embedded in the social conventions of the community, a period of great difficulty might well ensue. The new equilibrium total of population at Z would be at a lower standard of living than the old because of the changed fertility schedule. Figure 6.7, taken in conjunction with Figure 6.6 shows how this could come about. The new total of population is much larger than the old, and although supported by a much larger agricultural output, is unable to sustain the living standards which once existed. There might well be danger that the more vulnerable soils (perhaps in the older chalk areas) would be used too intensively under the new pressure of population and become permanently impoverished. The society might then learn once more to adopt social conventions which would result in a lower fertility schedule; or, failing that, become inured to a lower long-term standard of living (the history of Irish population between, say, 1750 and 1900 is extremely instructive in this connection – see Connell, 1950 and 1962; and Drake, 1963).

This illustration might be much more extensively developed and modified but is meant only to establish one point – that a new uncertainty and dynamism is introduced into the situation once the idea of a flexible ceiling is introduced into the model. Indeed the ramifications of this may be very intricate since it is perfectly possible that, given changes of the sort just described, new advances in material culture would be thrown up in response to the growing adversity in which the society found itself at the end of a period of easy expansion. For example, the next important advance might be towards a more intensive agriculture rather than an extensive expansion of the sort experienced in the preceding phase.

The fact that since the Neolithic food revolution changes in the material culture of societies have been frequent has destroyed the simplicity of the model appropriate for a society of static material culture. Both in the study of the developing areas of the world today and in the conventional topics of historical geography the nature of the material culture has rightly come to occupy much attention (see Wrigley, 1965A). It is the dynamic factor in so many situations. Yet the importance of understanding the range of demographic possibilities before a community is not affected by this except in the sense that it is much less easy to deal with the matter briefly and categorically. It is still true that the intensity of land use, the extent to which unpromising land is used, the division of the land between plough and pasture, the degree of subdivision of the land, and so on, are all deeply affected by the interplay between the fertility, mortality and nuptiality of the community. The English peasants of the thirteenth century were pushed closer and closer to the edge of a Malthusian precipice by their demography. Those of the fifteenth century lived better, used the land in different ways (relatively much more pasture – see Postan, 1962), and held the land upon different conditions, in part at least because the demography of the countrymen of Henry V was quite different from that of their forebears 150 years earlier. This state of affairs was not the temporary aftermath of the Black Death but continued for more than a century thereafter. That it later gave way to renewed population pressure and another cycle of events reminiscent in some ways of the thirteenth century does not alter the fact that the demographic situation for a century or more was sufficiently different from what it had been before or was to become later to have a great influence on living standards and land use.

Similarly, it is possible that demography had something to do with the contrasts which existed between different parts of south-east Asia in the period before the recent drastic fall in mortality. There were substantial differences in production per head and in real incomes in this area in spite of the fact that in many instances agricultural techniques, food crops and even environmental conditions were broadly similar. If the populations had all been characterized by the same fertility and mortality schedules it would be reasonable to expect that real income, for example, would have been at much the same level in areas which had much in common in other respects (provided always that there had not been major recent changes in material technology). Indeed it would not be absurd to claim that given similar fertility and mortality schedules it would be largely a matter of indifference whether an area was very fertile or largely barren, whether the climate was favourable to the growth of crops or marginal, since in the course of time the relative pressure of population upon resources would become much the same. Where, however, fertility and mortality schedules vary considerably, there is much greater scope for regional diversity. This is one possible explanation of those situations where peasants living on the richest land prove less wealthy on an average than others living in more adverse conditions of soil or climate. If the demographic characteristics of the latter are such that the equilibrium population total in the poorer areas is close to the optimum attainable with existing techniques, whereas in the former the point of balance between births and deaths is nearer to the maximum, those living in the first area may be less well off than those in the second.[1]

In considering the possible effects of different combinations of fertility and mortality schedules upon the long-term tendency of a pre-industrial population to stabilize in numbers at a particular level two further points should be borne in mind. The first is that what has been written, in addition to having the drawbacks of any schematic model, is likely to prove true only in secular tendency. It was a very common, perhaps a universal, feature of pre-industrial societies that in the shorter term there were very large fluctuations in births and deaths and to a lesser extent in population totals as a result of the ravages of famine and disease. A roughly cyclical movement of population totals, often with a periodicity of about thirty years, has been found in many parts of pre-industrial Europe, for example. A typical sequence of events was a bad harvest, or still more two bad harvests on the run, leading to undernourishment and accompanied or followed by serious epidemic infections. Such a crisis often reduced populations by a fifth or a quarter (see especially Goubert 1960: also Drake, 1964). There might then follow a generation of recovery with a surplus of births over deaths in most years before the abrupt downswing in the cycle again took place. Cycles of this sort might occur whatever the relative pressure of population upon resources, but would be much more severe in areas where the long-run equilibrium total of population implied by the fertility and mortality schedules was close to the maximum.

Secondly, it is important to be aware of the secondary effects of the demographic characteristics of a given population. For example, the age composition of a population is determined by its fertility and mortality schedules (for the sake of simplicity of exposition, I ignore migration, though it is often, of course, of great importance). This may be a fact of some importance in helping to understand living standards and patterns of expenditure, which in turn have an influence on land use and the structure of demand for industrial products. The specimen stable population figures published by the United Nations (United Nations, 1956B, pp. 26–27) may be used to illustrate this. These show that in a population in which the gross reproduction rate is 2·5 and the expectation of life at birth is 30 years the percentage of the total population between 15 and 59 years will be 57·6 per cent. If fertility remains unchanged but expectation of life at birth improves to 60 years the percentage of people in this age range falls to 52·6 per cent. These figures are for stable populations, that is populations in which the implications of fertility and

---

[1] The opposite may also occur, of course. There are many examples of poor agricultural or pastoral areas which suffered from rural under-employment, developing rural industries to occupy idle hands, encouraging thereby a further growth of population, and ending with a large but miserably poor population (see, for example, Deprez, 1965A and 1965B; and also an interesting discussion of the same general issue in Thirsk, 1961).

mortality schedules have been fully worked out. Any change in mortality as drastic as that just mentioned would mean that several generations must elapse before the new stable age structure was established. Nevertheless it represents the type of change which has taken place recently in many developing countries. If their present high fertility levels do not fall and mortality continues to improve the dependent 'tail' in the population must grow. If the income of the breadwinners does not rise, they will be hard pressed to provide for their more numerous dependants. More and more of the total demand for goods will be devoted to the basic necessities of life, and above all food. If population pressure is in any case intense the situation will be aggravated by this unfavourable development in age structure. If, on the other hand, the fall in mortality is soon followed by a decline in fertility sufficient to reduce the gross reproduction rate from 2·5 to 1·5, then the percentage of population 15–59 in the stable population which would ultimately develop would be very close to the original figure, but now there would be a far higher percentage of people of 60 or more and a lower percentage under 15 (in the first case 36·9 per cent 0–14, 57·6 per cent 15–59, 5·5 per cent 60 plus: in the second case, 28·2, 58·7 and 13·1 per cent respectively).[1]

Again, within any pre-industrial population there will probably be important differences between one region and another, and between town and country. It is useful to bear it in mind, for example, that in addition to the economic problems of maintaining large cities in pre-industrial times, associated with small and variable food surpluses in rural areas and the difficulties of moving food from country to town, there were also important demographic checks upon unrestricted town growth. If, as was frequently the case, mortality in the towns was at such a high level that the urban populations were only maintained by a constant flow of immigrants from the countryside, it is clear that if urban populations had risen above a certain fraction of the whole their birth deficit would have outweighed any surpluses occurring elsewhere and would have caused total population to fall. Pre-industrial urban populations were probably seldom large enough to lend much substance to this point, though it may turn out to be of importance in considering the relationship between London and the rest of England in the late seventeenth and early eighteenth centuries, when London was growing quickly, but the country as a whole grew little if at all.

More generally it is useful to know something of the results of different combinations of fertility and mortality on rates of growth of total population. This is particularly important in any topic related to the contemporary growth of population in the developing countries, sometimes referred to as the 'population explosion'. However steep the fall in mortality in these countries

---

[1] It may be noted that where a population is both stable and unchanging in total number, the percentage of the population of working age, say 15–59, is remarkably constant, whatever the combination of fertility and mortality which brings this about.

and however slight the changes in fertility, rates of growth of population will almost certainly never exceed five per cent per annum, and are very unlikely to reach even four per cent. Indeed rates much in excess of three per cent are uncommon. These are still formidable figures (three per cent per annum means a doubling of numbers in 23 years), but it is a fortunate circumstance that human physiology, unlike that of most animal species, does not permit faster rates of growth. Otherwise the economic prospects of some parts of Asia, Africa and South America today might be poor indeed.

DEMOGRAPHIC MODELS AND POST-INDUSTRIAL SOCIETIES

I have now touched upon the implications of demographic models when the material culture may be regarded as unchanging, as with Australian aborigines or South African Bushmen; and have broadened the discussion to enable agricultural economies to be considered also, those economies in which an advance in the techniques of cultivation, or transport, or civil administration may change the level of the ceiling of population and so permit renewed growth, perhaps with a new set of fertility and mortality schedules. Further modifications to the simplicity of the original model are necessary to embrace post-industrial economies. Whereas in hunting and fishing economies one may posit an unchanging ceiling, and in agricultural societies a ceiling which may inch up gradually or even rise with a jerk from time to time but which is still a viable concept for certain analytic purposes (it was, after all, of such a society that Malthus himself wrote), the very concept of a ceiling to population growth is inappropriate to a country which has undergone a full Industrial Revolution. In industrialized countries today indeed it is very difficult to give much precision to the idea of an optimum or maximum population. Certainly it is true to say that the criteria which can best be used in the discussion of this issue are no longer closely tied to the food base and the land. Populations have almost everywhere risen immensely in the last two centuries but living standards have also risen fast. Mortality, especially in the younger age-groups, has fallen precipitately. Exogenous mortality[1] has ceased to affect general mortality significantly. Fertility has also fallen very substantially and family size has become a matter of conscious individual choice to a degree which has few if any parallels in pre-industrial societies.

---

[1] Exogenous mortality consists of deaths from infectious diseases such as tuberculosis, smallpox, dysentery, typhus, malaria and plague, which caused immense loss of life until recently. Endogenous mortality, on the other hand, is made up of deaths from circulatory defects, organic malfunction, and (perhaps) cancer, which are not produced by the entry into the body of harmful viruses and bacteria.

In industrialized countries the vast majority of people live in cities: only a tiny fraction still make a living from the land. In these circumstances the mortality and fertility schedules need not change significantly as population rises, nor if they change are the changes necessarily in conformity with what might be expected on the assumptions embodied in the earlier models. Increasing density of population need not mean a rise in mortality. Changes in fertility may occur in either direction. Fertility and mortality schedules in which fertility and mortality levels are plotted against population density might take the form of a horizontal straight line as plausibly as any other, at least within the range of densities known today. It no longer makes sense even as a theoretical exercise to construct fertility and mortality schedules of the type to be found in the figures earlier in this chapter in which the shape of the curves is directly related to population size. If plotted against time rather than population density the mortality schedule might show a small decline as the years pass, while the fertility schedule would be indeterminate on present evidence. Changes in material culture which in very primitive societies may be so slight over long periods as to be negligible, and which in pre-industrial societies which cultivated the land, though swifter, were still halting and slow, have now become so sweeping and frequent that models which depend upon the absence of insignificance of these changes are of no utility. Nor is the level of real income per head closely connected with population density in the manner familiar in pre-industrial societies. Demographic model building for the study of contemporary problems must therefore take a different form.

In industrialized countries very little life is lost before men and women have passed through the reproductive ages. No foreseeable change in mortality rates, therefore, can greatly change rates of population growth and attention is naturally concentrated on fertility, except in the study of certain special problems. If, for example, current fertility levels in England are maintained till the end of the century it is possible to predict within small margins of error (largely the product of uncertainty about migration) how large the population in the year 2000 will be, not only in total but by age-groups. Mortality changes will probably be very small and can be largely ignored, saving a nuclear disaster or other unpredictable catastrophe. In any pre-industrial society, harassed by disease and famine, to ignore possible changes in mortality in this way would have meant inviting large errors in forecasting. He would be a foolhardy man who placed much store upon the continuance of present fertility rates for a period of more than a generation, but by making a series of alternative assumptions about fertility (and migration and mortality, of course, though the second is of slight importance) a range of possible population sizes can be ascertained. This in turn can be coupled to other assumptions; for example about the ratio between school-teachers and children of school age. Estimates of future needs of this sort cannot be very exact because several assumptions are involved and errors may be compounded. The number of schoolteachers needed is a function not merely of the number of children to be taught but also of average class size, and so on. But population projections are of importance none the less. Reasoned estimates are much more valuable than intuitive hunches. The simple models of demographic change upon which estimates of this sort are based are of great importance to those geographers interested in the industrialized countries (see, for example, United Nations, 1956A). Studies of resource use, of changing land use, of water conservation, of urban growth, of new patterns of population distribution, of regional growth and decay, and of migration, all require a knowledge of the population movements of the immediate past, of the present demographic situation, and of the changes which are likely in the near future. If population in general can reasonably be thought of as central to the whole range of issues with which human geographers occupy themselves (see Wrigley, 1965B) then a knowledge of demography can hardly fail to be of importance also.

THE DEVELOPING COUNTRIES

The case for an understanding of demographic models needs least advocacy perhaps in the case of the countries of Asia, Africa and Latin America which are nowadays usually referred to as the developing countries. Many of them are placed in a position of grave difficulty by their present demographic constitution. The nature of the problem is well known. In these areas in the last generation or so mortality has fallen precipitately because of the use of modern public health measures and powerful drugs. Fertility has changed little. In these circumstances population increase at the rate of two or even three per cent per annum frequently occurs. If the economies of these countries had changed as completely as their mortality rates there would not necessarily be any cause for alarm in these developments. Unfortunately the changes have not always been accompanied by rapid economic growth. Whereas in Europe the mortality changes which occurred in the early decades of the Industrial Revolution down to the end of the nineteenth century were slow and were associated with economic growth rather than medical advance, in the developing countries today this is not the case. In them health and wealth have not always moved in step.

The danger is obvious. If a drastic fall in mortality occurs without any change in fertility and without revolutionary changes in the economy, the ultimate adjustment may be exceedingly severe. Mortality in these circumstances remains ultimately density-dependent in the classical manner and the mortality schedule must turn up very sharply if population continues to grow. The gloomiest fears have been entertained on this score. There are good reasons for supposing that the worst of these fears will prove ground-less,[1] but in the meanwhile the life of these countries is dominated by the demographic situation. In Egypt, Indonesia and Ceylon today, for example, it must be the starting point for any adequate discussion of economy, society and human geography, and the implications of the current state of affairs can only be set out adequately in the light of a knowledge of the functioning of demographic models. For example, the population of Ceylon rose by 31·2 per cent in the years 1945–55; that of Japan by 23·3 per cent in the same period (United Nations, 1963, pp. 154–155). It might seem that in these two countries, both densely populated and both experiencing a rather rapid rise in popula-tion, the population problems were broadly similar. Yet demographically they had little in common apart from the rate of increase, for the rise in Japanese population in that period was largely the result of the age composi-tion of the population and the repatriation of Japanese nationals after the war. The net reproduction rate was not very much above unity and was falling – that is to say that if the prevailing trends in fertility and mortality had con-tinued population must soon have ceased to grow quickly. In Ceylon, on the other hand, the demographic situation was quite different. Any forecast of future population based on this knowledge was bound to include a con-siderable period of time during which the rate of growth would increase. Only a substantial and immediate fall in Ceylonese fertility rates could have falsified such a prediction.

The general possibilities of economic growth in countries like Ceylon is closely bound up with demographic developments. Each government in attempting to foster an economic 'take-off' is obliged to wrestle with the tendency for the rate of growth of population to accelerate. The Indian planners for more than a decade have been caught in the same predicament as the Red Queen. They must always run faster than they were running a little earlier simply in order to keep up. If they fail to keep the economy growing

[1] Reliable fertility statistics are few and far between for the developing countries. Un-doubtedly in many of them fertility has not fallen at all as yet and may even in some cases be rising. But the crude birth rates for 1959–64 for Formosa, Hong Kong and Singapore do suggest the possibility of rapid change in favourable circumstances. Hong Kong and Singapore are, of course, special cases, but the Formosan figures, since they refer to a population with a large rural component, may be very significant.

Crude birth-rates per thousand total population

	1959	1960	1961	1962	1963	1964
Formosa	41·2	39·5	38·3	37·4	36·3	34·5
Hong Kong	35·2	36·0	34·2	32·8	32·1	29·4
Singapore	40·3	38·7	36·5	35·1	34·7	32·1

Source: *United Nations Monthly Bulletin of Statistics*, July 1965, p. 5.

more quickly than the population real incomes per head will decline, demand for the consumer goods produced by industry may be expected to fall, capital saving to become still more difficult, and so on. The chances of success grow slim. Coale and Hoover's analysis of India's economic prospects (Coale and Hoover, 1958: see also Schwartzberg, 1963) gains greatly in depth and subtlety because of their care to consider a number of possible courses of demographic change and to show how these might affect economic growth in India.

General economico-demographic models can generate a wide range of important results. For example, in order to relate future increases of population in India to the degree of pressure on the land, it is not enough simply to be aware of the probable increase in total population over the next decade or quarter century. What may be called the secondary as well as the primary material produced by models of this type is useful. It would be naïve to suppose that a population increase of 50 per cent over the next 25 years would necessarily involve an increase in rural population of the same order of magnitude. If real incomes per head did not increase this might prove a valid assumption, with all that it implies for further subdivision of holdings and rural underemployment. If, on the other hand, the model shows that this rise in population is consonant with rising real incomes, the percentage of people living on the land will fall. Indeed a sufficiently steep rise in real incomes might imply that the absolute numbers of the rural population would stabilize within this period, or even decline. It would probably emerge from the consideration of alternative possible rates of growth that the faster the rate of increase, the slower the decline of the percentage of the population directly dependent on the land for income. Since the highest percentages on the land might be associated with the largest absolute population totals, the range of possible rural population totals would be wide. This is an issue of great significance for many aspects of social and economic life in developing countries. The merits of the assumptions built into any one model used in this connection must be a subject for dispute. But the general merits of tackling the problem in this way seem clear. The equations expressing the relationship between rates of population growth, trends in income per head, the structure of demand, and the fraction of the labour force employed on the land, may be mistaken, but at least they make it possible to explore the effect of changes in one or more variables upon the other variables in the system in a coherent and logical fashion. No human geography of an area of this sort which seeks to analyse as well as describe can afford to neglect the opportunities offered by models either of the narrow or the broader type.

It may be appropriate to note further here that models expressing the relationship between the demography of a society and other economic, social and geographical variables are inevitably of importance in studies of Industrial Revolution and the 'take-off' generally. For example, if it is agreed that at least until the 1780's the growth of industrial production in England was sustained largely by the strength of home demand, then it follows that the demographic constitution of the country must have embodied some unexpected features. The growth of home demand was comparatively slow and related to slowly rising real incomes. If Malthus had been right in supposing that the mass of the people would marry earlier and bring up larger families when economic circumstances were favourable, it is hard to imagine real incomes rising steadily for several successive generations. If they had behaved as he supposed (and there is much evidence suggesting that his model holds good for, say, the sixteenth century and at other earlier periods), events could not have fallen out as they did (see Wrigley, 1966 for a fuller discussion of this issue). More generally, in as much as the level of real incomes per head is important in studies of this type, demography must also be important since population size must enter into any calculation of income per head as one half of the ratio.

CONCLUSION

In describing and discussing where people live, how they obtain a living, and in what numbers they are to be found, geographers have become more and more sophisticated as the years have passed. The attempt to deal with these issues by referring aspects of population density and distribution directly to features of the physical environment was seldom considered satisfactory even in the infancy of modern human geography. Models of economic activity and of general spatial organization have been called into play increasingly in recent years to supersede or supplement older ideas, especially when dealing with areas which have experienced an Industrial Revolution. Sociological models and models of urban function have made an appearance (some are dealt with in Chaps. 7 and 9). Models which are derived from the consideration of formal geometrical properties are invoked to foster the understanding of the development of route networks and other problems. And many others might be added to the list. The use of models drawn from so many different sources, while it greatly increases the range of conceptual tools available, brings in its train other difficulties. For example, it is desirable that all should be subsumed under some overarching general model so that what each implies for the others can be examined (hence the great influence which the writings of August Lösch have exerted on geographers of the last two decades: his was a brave attempt at an important general solution). An ideal general model should make it possible to examine the effect which change in any one aspect of the segment of reality subsumed within the model will have upon all others.

The justification for introducing demographic models into the normal ambience of analysis and thus complicating matters still further lies partly in the rapidly growing body of empirical evidence which shows the importance of demography to many points of interest to geographers, and partly in the rapid improvement in the demographic models themselves in recent years. The Population Branch of the United Nations Bureau of Social Affairs has played a notable part since the war in promoting work of both types. Equally notable has been the work on theoretical, historical and contemporary demography carried out at the *Institut National d'Etudes Démographiques* in Paris. As a result of work done under the aegis of these two bodies and by a host of individual scholars there is ample evidence available both from empirical studies and theoretical works that it is impossible to treat population as a simple, dependent variable. Populations have, as it were, a life of their own, in part a reflection of their social and physical environments, perhaps in an absolute sense wholly their product, but in any proximate analysis best considered as a partially independent variable which is capable of modifying profoundly many aspects of economy, society and geography. To neglect demography is to shut oneself off from one of the most important sources of insight into the question of where people live, in what numbers and by what means.

REFERENCES

Note: this bibliography is restricted largely to works referred to in the chapter but includes a few other works of general usefulness.

ACKERMAN, E. A., [1959], Geography and demography; In *The Study of Population*, (Ed. Hauser, P. M. and Duncan, O. D.), (Chicago), 717–727.

BEAUJEU-GARNIER, J., [1956–58], *Géographie de la population*; 2 vols, (Paris).

BEAUJEU-GARNIER, J., [1966], *Geography of Population*, (London).

BOSERUP, E., [1965], *The Conditions of Agricultural Growth*.

CARR-SAUNDERS, A. M., [1922], *The Population Problem*, (Oxford).

COALE, A. J. and HOOVER, E. M., [1958], *Population Growth and Economic Development in Low-income Countries: a case study of India's prospects*, (Princeton).

CONNELL, K. H., [1950], *The Population of Ireland 1750–1845*, (Oxford).

CONNELL, K. H., [1962], Peasant marriage in Ireland: its structure and development since the famine; *Economic History Review*, 2nd ser., xiv, 502–523.

COX, P. R., [1959], *Demography*; 3rd ed., (Cambridge).

DARWIN, C., [1869], *On the Origin of Species*; 5th ed., (London).

DARWIN, C., [1901], *The Descent of Man*; new ed., (London).

DARWIN, C., [1958], *Essay of 1844*; In *Charles Darwin and Alfred Russel Wallace. Evolution by natural selection*, Pub. for XV International Congress of Zoology and the Linnean Society of London, (Cambridge).

DENIEL, R. and HENRY, L., [1965], La population d'un village du Nord de la France, Sainghin-en-Mélantois, de 1665 à 1851; *Population*, 563–602.

DEPREZ, P., [1965A], The demographic development of Flanders in the eighteenth century; In *Population in History* (Ed. Glass, D. V. and Eversley, D. E. C.), (London), 608–630.

DEPREZ, P., [1965B], *Evolution démographique et évolution économique en Flandre au 18e siecle*; mimeographed paper for Section VII, Third International Economic History Conference, (Münich), 1965.

DORJAHN, V. R., [1958], Fertility, polygyny and their interrelations in Temne society; *American Anthropologist*, 60, 838–860.

DRAKE, K. M., [1963], Marriage and population growth in Ireland, 1750–1845; *Economic History Review*, 2nd ser., xvi, 301–313.

DRAKE, K. M., [1964], *Marriage and Population Growth in Norway, 1735–1865*; unpub. Ph.D. thesis (Cambridge).

GLASS, D. V. and EVERSLEY, D. E. C., (Eds.), [1965], *Population in History*, (London).

GOODE, W. J., [1963], *World revolution and family patterns*, (New York).

GOUBERT, P., [1960], *Beauvais et le Beauvaisis de 1600 à 1730*; S.E.V.P.E.N.

HAJNAL, J., [1965], European marriage patterns in perspective; in *Population in history* (Ed. Glass, D.V. and Eversley, D. E. C.), (London), 101–143.

HATT, P. K., [1952], *Backgrounds of Human Fertility in Puerto Rico*, (Princeton).

HAUSER, P. M. and DUNCAN, O. D. (Eds.), [1959], *The Study of Population*, (Chicago).

HERLIHY, D., [1965], Population, plague and social change in rural Pistoia 1201–1430; *Economic History Review*, 2nd ser., xviii, 225–244.

HIMES, N. E.,]1936], *Medical History of Contraception*, (Baltimore).

HOLLINGSWORTH, T. H., [1964], The demography of the British peerage; supplement to *Population Studies*, xviii.

JAFFE, A. J., [1951], *Handbook of Statistical Methods for Demographers*; (Washington).

KRZYWICKI, L., [1934], *Primitive Society and its Vital Statistics*, (London).

LEIBENSTEIN, H., [1954], *A Theory of Economic-Demographic Development*, (Princeton).

LEVY, M. J. and WESTOFF, C. F., [1965], Simulation of kinship systems; *New Scientist*, xxvii, No. 459, 571–572.

LORIMER, F. and others, [1954], *Culture and Human Fertility*, (UNESCO).

LOTKA, A. J., [1934–39], *Théorie analytique des associations biologiques*; 2 parts, (Paris).

LUND STUDIES IN GEOGRAPHY, [1957], Ser. B, Human Geography, No. 13, *Migration in Sweden* (Lund).

POSTAN, M. M., [1950], Some economic evidence of declining population in the later Middle Ages; *Economic History Review*, 2nd ser., ii, 221–246.

POSTAN, M. M., [1962], Village livestock in the thirteenth century; *Economic History Review*, 2nd ser., xv, 219–249.

POSTAN, M. M. and TITOW, J., [1959], Heriots and prices on Winchester estates; *Economic History Review*, 2nd ser., xii, 392–417.

PRESSAT, R., [1961], *L'Analyse démographique*, (Paris).

SAUVY, A., [1952–54], *Théorie générale de la population*; 2 vols., (Paris).

SCHWARTZBERG, J. E., [1963], Agricultural labour in India: a regional analysis with particular reference to population growth; *Economic Development and Cultural Change*, xi, 337–352.

SPENCER, B. and GILLEN, F. J., [1927], *The Arunta: a study of a Stone Age people*; 2 vols, (London).

THIRSK, J., [1961], Industries in the countryside; In *Essays in the Economic and Social History of Tudor and Stuart England* (Ed. Fisher, F. J.), (Cambridge), 70–88.

UNITED NATIONS, [1949], Population Branch of the Department of Social Affairs, Report No. 7, *Methods of using census statistics*.

UNITED NATIONS, [1953], Population Branch of the Department of Social Affairs, Report No. 17, *The determinants and consequences of population trends*.

UNITED NATIONS, [1956A], Population Branch of the Department of Social Affairs, Report No. 25, *Methods for population projections by sex and age*.

UNITED NATIONS, [1956B], Population Branch of the Department of Social Affairs, Report No. 26, *The aging of populations and its economic and social implications*.

UNITED NATIONS, [1963], *Demographic Yearbook*.

WATT, K. E. F., [1962], The effect of population density on fecundity in insects; *General Systems Yearbook*, 7, 231–244.

WESTERMARCK, E., [1921], *The History of Human Marriage*; 5th edn., 3 vols., (London).

WOLFENDEN, H. H., [1954], *Population Statistics and their Compilation*; 2nd edn., (Chicago).

WRIGLEY, E. A., [1962], The supply of raw materials in the Industrial Revolution; *Economic History Review*, 2nd ser., xv, 1–16.

WRIGLEY, E. A., [1965A], Changes in the philosophy of geography; In *Frontiers of Geographical Teaching*, (Ed. Chorley, R. J. and Haggett, P.), (London).

WRIGLEY, E. A., [1965B], Geography and population; In *Frontiers of Geographical Teaching*, (Ed. Chorley, R. J. and Haggett, P.), (London).

WRIGLEY, E. A., [1966], Family limitation in pre-industrial England; *Economic History Review*, 2nd ser., xviii, 82–109.

WYNNE-EDWARDS, V. C., [1962], *Animal Dispersion in Relation to Social Behaviour*, (Edinburgh and London).

CHAPTER SEVEN

Sociological Models in Geography[1]

R. E. PAHL

The Poet Wonders Whether the Course of Human History is a Progress, a Drama, a Retrogression, a Cycle, an Undulation, a Vortex, a Right- or Left-Handed Spiral, a Mere Continuum, or What Have You. Certain Evidence is Brought Forward, but of an Ambiguous and Inconclusive Nature. (Title to Ch. 18 of *The Sot Weed Factor* by JOHN BARTH)

The important question for the sociologist is not whether he should interpret observed human behaviour in terms of models, but what sort of model he should employ. (JOHN REX)

Our understanding of complex reality depends on the questions we ask: these in turn depend upon the culture into which we have been socialized and the viewpoint or discipline in which we have been trained. A fact is only a fact in terms of a theory and problems are defined as such within an implicit or explicit theoretical framework. That such conceptual schemes are more often only tacitly accepted in empirical investigation should be a matter of deep concern to social scientists. It seems clear that any attempt to explain the ordering and pattern of human activity is constricted if forced into a mono-causal framework. Geographers in particular, who may attempt to 'explain' medieval field-systems or economic growth in modernizing countries by 'the facts' should see the need for a more rigorous theoretical framework. Seemingly every generation is obliged to rewrite the past in terms of its own values and ideologies. Yet however strongly we may react against a naïve historicism it cannot be denied that an understanding of the past is internalized in the individual – particularly in certain charismatic leaders – and may be partly responsible for spatial and temporal diversity of organized human activity.

[1] I am grateful to Geoffrey Hawthorn of the Department of Sociology at the University of Essex and my fellow sociologist Rex Taylor at the University of Kent at Canterbury for helpful comments on this essay. We were all trained as geographers.

INTERNALIZED MODELS

Everyday behaviour is based on a generalized and structured concept of reality. We react to situations and persons in accordance with our stored experience; only in unusual situations without precedent do we find ourselves at a loss, that is, when our mental models' approximation to reality is slight. A resistance to the acceptance of alternative models at the individual level is normal; at a more general level such resistance may take the form of ethnocentrism, which clearly demonstrates, on a comparative basis, the nature of a culturally bestowed model. It would be interesting to consider the degree of ethnocentricity as between nations and also between groups within nations. Certainly it is difficult to be objective about one's own cultural or sub-cultural values. Academics, who might be expected to be more sympathetic to the relativity of their socially bestowed attitudes, and who may discuss with considerable sophistication the cultural differences between, say, the French peasant and the English manual worker, may be surprised and somewhat shocked if a son of the latter has difficulty in accepting the different values of what might be a middle-class dominated residential university.

If we consider the internalized models we may have of the British nation state, it is possible that collectively the nation is suffering from the delay in rejecting an outdated model of Britain as an Imperial power. The newly independent states, once part of the British Empire, have in many cases rejected some of the distinctive cultural and political values of the administrative elite, often partly formed by the academics who trained them. This attack on a model, which many assumed did not need to be defended, is paralleled by the widespread acceptance of some form of socialism over most of the world for most people, leaving a much diluted form of capitalism to a geographically small but economically rich area in Europe and North America, with isolated outposts in Australasia and Japan. This double attack has had the effect of administering something of a national shock, which needs more than the growth of coffee drinking and central heating over the past fifteen years to cure – however much a hot drink and keeping warm may be the traditional cures for shock.

These attacks on our internalized models of the nature of our own nation in relation to the rest of the world, and the attempts to come to terms with the changed situation, have had repercussions on our whole intellectual climate and throughout our educational system. It is therefore not surprising that in this situation those responsible for teaching and research in geography should question some of the basic assumptions which had become fossilized in the conventional wisdom of the textbooks. Philip Abrams' strictures on the teaching of history are equally applicable to the teaching of geography – 'Sometimes one feels that . . . teachers and authors are incapable of recognizing a controversial subject when they see one – the number of fourth- and fifth-form histories that treat unions as things that simply "grew" or slumps as things that simply "set in" is truly remarkable.' Similarly in geography controversial aspects of economic development and regional planning are played down. Alternative models to explain, say, the local climate are not always matched by a similar sophistication when considering other aspects of economic development. One textbook will have to suffice as an example – *The Mediterranean Lands* by D. S. Walker, which just happened to be the first that came to hand.

In his discussion on the economic development of Spain, Walker states that 'The depression of 1929 and onwards coincided with a period of political experimentation which provoked such violent opposition that civil war broke out in 1936.' This led, we are told, to the near collapse of the Spanish economy. The Second World War prevented aid coming from abroad and 'Even after the end of hostilities she remained isolated economically and ostracized politically.' (The reasons for this are not explained.) Thus, Walker continues, 'In the light of these events the adoption by the present régime of an autarkic policy is understandable; they have made a virtue of necessity' (p. 110). And this is a sixth-form textbook! Turning to Portugal, where the main problem is said to be overpopulation, Walker notes: 'The traditional remedy is emigration and Portugal is fortunate in possessing outlets in Africa and cultural connexions with Brazil which make it possible for the emigrant to acclimatize himself rapidly' (p. 142). It appears that most emigrants prefer Brazil 'to the colonies'. (The reasons for this are not explained.) That the fortunate outlet could also be described as a repressively paternalistic forced labour state is not even hinted at, although James Duffy's *Portuguese Africa* was published in 1959, the year before Walker's book appeared.

THE MYTH OF A VALUE-FREE GEOGRAPHY

When geographers move away from the mapping of a static situation to consider aspects of change they inevitably become involved with values – both by the problems they choose to consider and the interpretation of the 'facts' they present. The interpretation of socialist economic development may vary according to the political sympathies of the geographer concerned: failure of the harvest in China might be held as an indictment of the political system, whereas a similar catastrophe in India may simply be seen as an Act of God. Clearly the issue is not as clear cut in actual practice, but there is need for considerable alertness in spotting the ideological bias. It is possible that the spatial element in a given situation may dictate action more forcibly in a capitalist society, whereas a socialist or centrally planned economy has a greater choice in the way resources are developed. Such greater choice would carry with it the possibility of greater error. When the state makes the wrong investment decision this is held to be an indictment of one system, yet when an individual firm goes bankrupt this is held to be a justification of the other. Certain geographers in France are held by some to be suspect because they are 'doctrinaire marxists'. This may be a justifiable accusation, but it would be nice to hear similar accusations of 'doctrinaire capitalists'. If we take as an example the problems of regional planning in Britain, it is clear that there can be strong divergence between those who project present trends, with some sort of historical inevitability, to predict the future, and those who instead propose alternative policies based on political interference with the so-called 'free play of market forces'. This distinction between 'drifters' and 'counter-drifters' is not always made explicit, although individual geographers have openly committed themselves by membership of technical advisory groups of all three political parties and there is by no means agreement by what criteria an abstract, 'ideal-type' geographer should advise the government of the day.

The geographer, as any other social scientist, cannot avoid being socialized into a specific culture – or, more accurately, sub-culture – at a particular period of time. Family, educational institutions, peer group, the mass media and so on have provided him with a certain model of the nature of society. By the material he chooses to teach from and the research problems he investigates, certain attitudes about the nature of society become incorporated into a pattern of thought, a system of values, which may or may not be made explicit. Whereas the relations between phenomena of the natural world can only be explained from the outside – and the geographical distribution of individuals and social groups can be observed in a sense as natural phenomena – relations between phenomena of the human world are relations of value and purpose.

SOCIOLOGICAL MODELS IN GEOGRAPHY

In the context of the other contributions to this symposium it is unnecessary to consider how models generally have been applied in sociology. That in itself is a fit subject for another symposium, aimed specifically at sociologists, who may be unfamiliar with the more refined mathematical techniques and their applicability to quantifiable data. Certainly the theory of games and other probabilistic models have considerable potential for developing certain branches of sociology, particularly those which view problems in a sub-social or materialistic way. It is perhaps significant that at the time of writing the most recent issue of the *European Journal of Sociology* is devoted to five articles on simulation models. Evaluation of the various elements in complex reality requires the use of the most rigorous and refined techniques available. The rather arid discussion about terminology – whether quantified theories should be called mathematical models (Brodbeck, 1959) and so on – need not detain us. The importance and relevance of such techniques in geography can be left to others in this book. Our concern here is with specifically *sociological* as opposed to mathematical or any other sort of models.

Although an interest in groups and 'societies' can, as with so many other interests, be traced back to the Greeks (Barnes and Becker, 1961), sociology is a very modern science. Social information does not constitute a science and sociology could not exist without someone first creating a model. 'The important question for the sociologist is not whether he should interpret observed human behaviour in terms of models, but what sort of model he should employ' (Rex, 1961, p. 60). However, it is one thing to discuss sociological models as such, about which sociologists may disagree amongst themselves, and it is quite another matter to relate such models to geography. Looking at the problems defined by one discipline with the conceptual tools of another is not easy within the existing intellectual climate, although this book is of itself a welcome sign of change. The social sciences have suffered individually from a parochial concern to defend the importance of the specific 'factors' with which they are most concerned. This 'fallacy of misplaced concreteness' has been attacked by Parsons – 'the effect of this tendency to "empirical closure" of a system is to make its application to any field, especially a new one, a rigidly simple question of whether it "applies" or not. Applications are interpreted in "all or none" terms – it is either a case or not' (1945, pp. 221–222). Factor theories are justifiable if they can demonstrate their effectiveness in solving empirical problems. Geographers would probably be the first to admit that in problems such as are involved in the economic development of some of the new states an understanding of the 'factors', such as the physical resource base or the spatial component, is of limited value. Similarly, economists accept the limitations of their more explicitly stated models. A crude reductionism heaps together perhaps the most significant elements in the situation in the geographer's 'other human factors' or the economist's escape clause 'other things being equal'. Both geographers and economists acknowledge the importance of the social factor in economic development but are not always clear how this can be understood in terms of some overall framework. Economists have different models which they apply in different situations – for example an economy of perfectly competitive enterprises (without monopoly), an economy of 'imperfect' competition – that is an economy containing some monopolistic power – or a socialist economy. They do not expect any one model to fit an actual society exactly. When we turn to sociology, as the author of a recent textbook puts it,

Instead of making various simplifying assumptions about social institutions and the distribution of economic power and then asking how certain variables are interrelated within such a purposely simplified framework, the sociologist focuses attention on the framework itself and asks what the institutional patterns are

3

within which economic activity is carried on, in what ways they are alike in all societies, in what ways they differ from one society to another and how a society comes to change its institutional pattern. (Johnson, 1960, p. 210.)

Indeed the concept of the 'economy' can be analysed in terms of a specific cally differentiated 'sub-system' of 'society' within the framework of a 'social system' (Parsons and Smelser, 1956). Social information thus gains considerably more significance in terms of such conceptual models as a social structure or a social system.

This should by no means be seen as an attack on geography as such, nor on any form of specialization, which, in view of the complexity of the subject matter of the social sciences, is both necessary and desirable. 'What is needed is a close collaboration between sociologists and other social scientists: and such collaboration implies both that the sociologist should have competence in one or other of the special social sciences, and that specialists should have some knowledge of general sociology' (Bottomore, 1962, p. 21). Clearly this is not the place to provide a geographer's introduction to sociology, although a book on this subject is certainly needed. Rather it is hoped that by discussing the sociologists' fundamental conceptual models, the geographer may gain a tool which may be of value in both formulating and clarifying his own empirical research, and which may also provide a framework into which other sociological work may be placed. It will be necessary to move to a fairly high level of abstraction in the discussion which follows. But it is perhaps useful at this point to pose the key question, which would presumably be of equal interest to both sociologists and geographers and on which we may test the conceptual models to be discussed: *Why are some societies different from others in the way they utilize their resources and distribute themselves in space, and, further, given a particular pattern of activities in relation to certain resources, what leads to change?*

As emerged above when discussing the role of values, it is the element of *change* in a situation which would appear to force the geographer away from a static, snap-shot approach. It is my view that mathematical and experimental models cannot be fully exploited in either geography or sociology until conceptual models of *changing* situations provide a better foundation from which to work. All social scientists talk about 'society' and yet its definition creates a host of problems, not to speak of the problem of what we mean when we talk of a society 'changing'. We turn now to a consideration of social structure and social system, outlining the way sociologists use the model of functional analysis, 'the most promising . . . of contemporary orientations to problems of sociological interpretations' (Merton, 1957, p. 19). In particular we consider the way this model stands up to the problem of change. Finally we consider the socio-ecological model in relation to the specific problem of evaluating the spatial component within a specific social structure.

THE NORMATIVE ORIENTATION OF ACTION SYSTEMS

Geographers, in so far as they take account of sociological theory, generally limit themselves to a demographic notion of social structure which of course lends itself more readily to the manipulation of data both cartographically and statistically. This is to impose an unnecessary restriction on the scope of the social sciences. Certainly it is possible to observe the external course of events with the accepted methods of natural science: elements of uniformity clearly do emerge in a study of social behaviour. But social scientists can, in addition, impute motives to men and 'interpret' their actions and words as expressions of their motives, for men act on the basis of *shared* values. A way of 'structuring social action' as a model, however rarified, is better than no model at all. The very important task of creating a specifically sociological frame of reference, not tied to environmentalism or psychological reduction-ism was greatly aided by Talcott Parsons in his somewhat involved but greatly influential work *The Structure of Social Action*, first published in 1937. The argument that the 'irrationality' of human action was simply due to the freedom of the will and therefore was not capable of scientific investigation (without calling on the laws of chance) was refuted by Max Weber. He pointed out that the sense of freedom would, if this were true, be associated primarily with irrational actions – that is those involving emotional outbreaks and so on – but in fact the reverse is more nearly true. It is when we act most rationally that we feel most free, not constrained by emotional elements. If we then accept that *rational action* is to a high degree both predictable and subject to analysis in terms of general concepts, it is of crucial importance to develop such concepts – which would clearly have relevance to all the social sciences.

'The normative orientation of action systems' can be a difficult concept to grasp because of the abstract nature of the concept and its somewhat tenuous connection with empirical investigations. Indeed, Parsons himself admits 'these concepts contain an element of "unreality" which is not involved in the physical sciences. Of course the only reason for admitting such concepts to a scientific theory is that they are in fact descriptive of an empirical phenome-non, namely the state of mind of the actor. They exist in this state of mind but not in the actor's external world' (Parsons, 1937, p. 295). In developing his voluntaristic theory of action Parsons refutes positivistic thought, caught in what he calls 'the utilitarian dilemma':

> That is, either the active agency of the actor in the choice of ends is an independent factor in action, and the end element must be random; or the objectionable impli-cations of the randomness of ends is denied, but then this independence disappears and they are assimilated to the conditions of the situation, that is to elements

analysable in terms of nonsubjective categories, principally heredity and environment in the analytical sense of biological theory. (Parsons, 1937, p. 64.)

Individualistic positivism is seen, perhaps, in its most acute form in the 'Hobbesian Problem of Order'. To Hobbes man is guided by a plurality of passions, desires are random and, since the ultimate ends of action are diverse, conflict would seemingly be inevitable. The war of all against all is avoided, according to Hobbes, by means of the idea of a social contract, whereby the actors come to realize the situation as a whole, instead of pursuing their own ends in terms of their own situation. Nevertheless, he saw that it is a direct corollary of the postulate of rationality that all men should seek and desire power over one another. 'Thus', as Parsons significantly remarks, 'the concept of power comes to occupy a central position in the analysis of the problem of order' (p. 93). His aim was to deal with this problem without making use of 'such an objectionable metaphysical prop as the doctrine of the natural identity of interests' (p. 102).

Parsons, in a masterly analysis of the works of Marshall, Pareto, Durkheim and Max Weber demonstrates the thread of a normative theory of action running through the work of each, despite the very different backgrounds and assumptions from which they start. Marshall rejected the idea of the egoism of the traditional economic man: he did not accept that society may be completely understood in terms of utilitarian want satisfaction but felt that it also involves certain common values. As Parsons put it 'economic actions . . . are also carried on for their own sake, they are modes of the immediate expression of ultimate value attitudes in action' (p. 167). Moving on to Pareto and Durkheim, Parsons argues that the former saw the individual integrated to some degree with others in a common value system, without seeing this 'social' element as a metaphysical entity in either a positivistic or an idealistic sense.

Durkheim went further, seeing the social factor as a system of ideas, which the actor passively contemplated. This led him on to identify the social factor with the *a priori* source of the categories, thus finally breaking the bond which had held it as part of empirical reality. Geographers interested in society could still read Durkheim with considerable profit – in particular his *Division of Labour in Society*. When discussing the incidence of suicide among Protestants and Catholics, Durkheim argues that the higher rate among the former is due to the different *content* of the different value systems: the individual responds to the norms and values of the group. It is easy to see from this sort of analysis how Durkheim came to see society 'out there' imposing itself on the individual. As Parsons puts it, in his evaluation of Durkheim's consideration of the social conditions of individual action:

Among these he found a crucial role to be played by a body of rules, independent of the immediate ends of action. In the end these rules are seen to be capable of interpretation as manifestations of the common value system of the community;

it is because of this that they are able to exercise moral authority over the individual. In so far as the immediate ends of particular acts are removed from ultimate ends by many links of the means – end chain, even though these ultimate ends be in conformity with the common system of ultimate values, there is need for a regulatory system of rules, explicit or implicit, legal or customary, which keeps action . . . in conformity with that system. The breakdown of this control is *anomie* or the war of all against all.

This body of rules governing action in pursuit of immediate ends in so far as they exercise moral authority derivable from a common value system may be called social institutions. (Parsons, 1937, p. 407.)

On two occasions Durkheim defined sociology as the science of institutions. In *The Elementary Forms of the Religious Life* Durkheim was forced further from any lingering positivism as being man's sole significant cognitive relation to external reality. Religion may be seen as one mode of human orientation towards the non-empirical: religious ideas are an important, and perhaps essential, element in the normative order. Durkheim did not exploit the fundamental point that Parsons draws from his work, namely that 'the central importance of religion lies in relation to action not to thought' (p. 441). The work of Max Weber, as we shall see, made this point explicit.

Durkheim found the essential element of order in common values, manifested above all in institutional norms, but he was primarily concerned with a stable system, more useful in the definition of *categories* of sociological analysis than the functional interrelations between them. He provided a basis on which a theory of social change could be based: such a theory could not emerge until it was known *what* changes. It was primarily through Durkheim's work that ends and norms were seen to be no longer individual but also social. The functional relationships between different elements in the system was a subject of deep concern to social anthropologists: before considering this functional model we may turn to the work of Max Weber.

WEBER'S 'IDEAL TYPE' MODELS

The relationship between the Protestant ethic and the spirit of capitalism, as discussed by Weber, is the work for which to non-sociologists he is best known. His comparative analyses of Confucianism, Buddhism, Hinduism and Judaism are not so well known and yet it is on the basis of this larger work that his work on the *Protestant Ethic* gains its force. Certainly it is difficult to provide any simple answer to the question which asks why, given the coal resources of Britain, India and China, those in Britain should be developed first. One can hardly postulate the environment or heredity form of single factor analysis, but it is also clear that Weber's work on ideas as causes and consequences, however strongly based on the most thorough and wide ranging scholarship, cannot provide any sort of holistic model of the processes of economic development. Nevertheless the analysis of Weber's work by Parsons (1937, pp. 500–578) or Bendix (1962), together, of course, with some basic elements of the original work (as, for example, available in Gerth and Mills, 1948, pp. 302–359) is essential reading for the social geographer interested in economic development, as a brilliant example of sociological analysis. There is, of course, no quantitative assessment of religion as a causative element in modern capitalism: it was a necessary though not sufficient condition. Weber's work, as far as this present chapter is concerned, is mainly of importance in illustrating the use of the concept of an 'ideal type'.

Weber outlined his methodical position in Chapter I of his posthumously published *'Wirtschaft und Gesellschaft'*; his remarks on ideal types are of particular interest:

> the same historical phenomenon may be in one aspect 'feudal', in another 'patrimonial', in another 'bureaucratic', and in still another 'charismatic'. In order to give a precise meaning to these terms, it is necessary for the sociologist to formulate pure ideal types of the corresponding forms of action which in each case involve the highest possible degree of logical integration by virtue of their complete adequacy on the level of meaning. But precisely because this is true, it is probably seldom if ever that a real phenomenon can be found which corresponds exactly to one of those ideally constructed pure types. The case is similar to a physical reaction which has been calculated on the assumption of an absolute vacuum. Theoretical analysis in the field of sociology is possibly only in terms of such pure types. . . . The more sharply and precisely the ideal type has been constructed, thus the more abstract and unrealistic in this sense it is, the better it is able to perform its methodological functions in formulating the clarification of terminology, and in the formulation of classifications, and of hypotheses. (Weber, 1947, pp. 110–111.)

Weber's tortuous prose hardly gains in clarity on translation and the fact that his work has had such a remarkable impact on the development of American sociology owes much to the penetration of his commentators. *The Theory of Economic and Social Organization*, which first appeared in translation in 1947, has a valuable introduction by Parsons. However, returning to the discussion of an ideal-type as an important sociological model, David Lockwood has provided a succinct summary of the characteristics of Weberian ideal-type analysis (Lockwood in Gould and Kolb, 1964, pp. 312–313), based on the passage from which we have just quoted. He notes that Weber's discussion of 'bureaucracy' is in terms of a true ideal-type since this phenomenon recurs in a variety of historical contexts (cf. Eisenstadt, 1963) and so it is truly a generalizing concept. However 'the Protestant Ethic' although discussed by Weber in terms of an ideal type can hardly be considered as such, since it lacks the abstract general quality of the former example. Further discussion of Weber and ideal-types by Carl Hempel and H. Stuart Hughes provides deeper analysis, and geographers as well as other social scientists may find them useful and thought-provoking accounts.

FUNCTIONALISM A MODEL

During the inter-war period when Parsons was working on his interpretation of Weber and moving towards his conception of a normative system of action, social anthropologists, through their work on primitive societies, were much concerned with the role specific forms of social activity played in the working of the whole society. Radcliffe-Brown, in 1935, discussed the analogy between social life and organic life, pointing out that any such analogy should be used with care. Like the life of an organism the continuity of a social structure is preserved through its functional continuity. As he put it:

– if we examine such a community as an African or Australian tribe we can recognize the existence of a social structure. Individual human beings, the essential units in this instance, are connected by a definite set of social relations into an integrated whole. The continuity of the social structure, like that of an organic structure, is not destroyed by changes in the units. Individuals may leave the society by death or otherwise: others may enter it. The continuity of structure is maintained by the process of social life, which consists of the activities and interactions of the human beings and of the organized groups into which they are united. The social life of the community is here defined as the *functioning* of the social structure. The function of any recurrent activity, such as the punishment of a crime, or a funeral ceremony, is the part it plays in the social life as a whole and therefore the contribution it makes to the maintenance of the social structure. (Radcliffe-Brown, 1952, p. 180.)

It is important to remember during the discussion which follows that Radcliffe-Brown made it quite clear that the idea of the functional unity of a social system was a *hypothesis*. He argued that in the same way that physics deals with the structure of atoms and colloidal chemistry with the structure of colloids, so also should there be a place for a branch of natural science dealing with the general characteristics of those social structures of which the component elements are human beings. The network of actually existing relations which makes up the social structure is, he argued, just as real as individual organisms.

The functionalism of social anthropologists has come under powerful attack, in particular from Rex and Dahrendorf. If we recall the question which we posed as basic to the argument of this chapter – the analysis of social change with a functionalist model – then we must consider what happens when a society falls into what Radcliffe-Brown calls 'a system of functional disunity or unconsistency'. The biological analogy suggest that ill-health or even death will follow in such a situation. It is easy to fall into the trap of regarding all changes as dysfunctional and of thinking that the old order represents a 'healthy' social organism which has to 'suffer' change. Certainly the analogy might be fruitfully pursued when considering recent events in the Congo, but it is difficult to apply it in the case of South Africa, where the social organism is seemingly maintained by the brute force of a minority. Clearly some things are more important than others – the body can still function without a limb but not without a heart or brain. As John Rex points out: 'The danger is that if the nature of the theoretical model which is unconsciously being used is not made explicit, the anthropologist is likely to interpret as 'functional' those activities which fit into his own scheme of goals and values, and what is more, represent them as essential to the survival of the society. It cannot be too strongly pointed out that explanations of social activities as performing functions in this sense carry no implication that without these activities the society would not survive. All that they imply is that without them certain goals would not be achieved' (Rex, 1961, p. 72).

The form of functionalism which has been developed by social anthropologists, based on a 'face-to-face' society where there are fairly clear-cut lines as between one society or tribe and another, is of less immediate relevance in the context of an advanced industrial society, where the important situations depend more on the relationships between institutions rather than between actors. Before making a more extended analysis of functionalism it will be useful to say something about *The Social System* by Talcott Parsons, itself a carefully worked out conceptual model.

Parsons aimed to create 'a body of logically interdependent generalized concepts of empirical reference' in such a state of logical purity that every logical implication of any combination of propositions in the system would be explicitly stated in some other proposition in the same system. He claimed that this would provide a 'genuinely technical analytical tool' which would avoid a situation in which vital elements were overlooked, and thus it would minimize 'the danger, so serious to common sense thinking, of filling gaps by resort to uncriticized residual categories'. (Parsons, 1954, p. 217.) This generalized theoretical conceptual framework he termed the structural-functional system (incidentally he also noted with approval the value of a physiological analogy).

Perhaps the most concise summary of Parsons' ideas appears in *Toward a General Theory of Action* (1951, esp. pp. 53–109) and those who wish to do full justice to them should read the original and not be misled by his 'translators' and commentators. 'Any behaviour of a living organism might be called action; but to be so called, it must be analysed in terms of the anticipated states of affairs toward which it is directed, the situation in which it occurs, the normative regulation (e.g. the intelligence) of the behaviour, and the expenditure of energy or 'motivation' involved. Behaviour which is reducible to these terms, then, is action' (p. 53). 'Actions are not empirically discrete but occur in constellations which we call systems. We are concerned with three systems, three modes of organization of the elements of action; those elements are organized as social systems, as personalities and as cultural systems' (p. 54). An analysis of cultural systems is essential to the theory of action 'because systems of value standards . . . and other patterns of culture, when *institutionalized* in social systems and *internalized* in personality systems, guide the actor with respect to both *the orientation to ends* and the *normative regulation* of means and of expressive activities' (p. 56).

The normative, ideal, aspects of the structure of systems of action, or the model of part of its culture are analysed by Parsons in terms of five pattern variables.[1] These are dichotomies 'one side of which must be chosen by an actor before the meaning of a situation is determinate for him, and thus before he can act with respect to that situation' (p. 77). The argument is that '*every* concrete need-disposition of personality, or every role-expectation of social structure involves a combination of values of the five pattern variables' (p. 93). Assuming that this list is exhaustive a cross-classification of each of the five against each of the others will provide a table of thirty-two cells and this can be seen as a first step towards the construction of a dynamic theory of systems of action. In fact, actuality is simpler than Parsons' theoretical complexities and there are empirical clusterings of the various structural components of society.

In so far as geographers are concerned with clarifying their conception of Parsons' notion of a social system there is much to be said for reading the original (Parsons, 1951, Chaps. 3, 4 and 5) although both Johnson (1961, Chap. 3) and Rex (1961, Chaps. 5 and 6) provide useful summaries. Parsons' social system is based on normative consensus: indeed he sees the integration of common value patterns with the internalized need – disposition structure of the actors involved as not only crucial for the stability of any social system but as the 'fundamental dynamic theorem of sociology' (1951, p. 42). This institutional integration may explain very little in detail but Parsons argues that his exposition, based on deductive reasoning, will enable empirical application of the conceptual scheme to follow. As he admits 'this is obviously a highly simplified model' (1951, p. 44) which hardly accommodates conflict and ambivalence in relation to the central value system of the society. 'Fulfilment of a given set of expectations will impose a greater 'strain' on one actor than another' (p. 45).

Unfortunately Parsons is unable to present a general theory of the processes of change of social systems, but he does concede the possibility of an 'increase of strains in one strategic area of the social structure which are

---

[1] The five pattern variables are: 1. Affectivity – Affective neutrality, 2. Self-Orientation – Collectivity orientation, 3. Universalism – Particularism, 4. Ascription – Achievement, 5. Specifity – Diffuseness.

finally resolved by a structural reorganization of the system' (p. 493). He feels that it is probably of more importance to trace the repercussions of change, once started, throughout the social system, including what he calls the 'backwash' of modification of the original direction of change (p. 494).

Following Johnson's admirable summary (1961, pp. 51–59) the main elements in the structure of a social system are:

1 Subgroups of various kinds, normatively related.
2 Roles both within the larger system and within subgroups – each role system being normatively related with each of the others.
3 Regulative norms governing subgroups and roles.
4 Cultural values.

Every social system must solve four functional problems:

1 Pattern maintenance and tension management.
2 Adaption
3 Goal attainment
4 Integration.

In very broad terms, the family, schools, religious groups and so on deal with the first problem, the economy deals with the adaptive subsystem of society, the polity or government deals with society's goals (although people may be more actively concerned with these goals at specific times) and the integrative subsystem is maintained by the legal profession, opinion formers of the mass media, religious leaders and so on. Each of these functional subsystems of a society can, in themselves, be analysed as a social system and the other systems are then seen as part of the 'environment' to which the subgroup must adapt if it is to survive and achieve its goals. Significantly Johnson comments 'It always happens that the cost of meeting societal problems is borne unequally by the subgroups of society. In our own society, for example, successful functioning of the economy requires that certain particular business firms must be allowed or 'forced' to fail; otherwise the economy as a whole would be less adaptive than it is' (p. 59).

FUNCTIONALISM AND THE PROBLEM OF CHANGE

There have been various hints in the preceding section suggesting some inadequacy in the functional model in coping with the problem of change. Much of the structural-functional model works simply by definition. For example, a given socio-technological system 'demands' an industrial organization of a certain degree of complexity within which, with the division of labour in society, actors perform the necessary roles to keep the enterprise functioning and others, in another sector of the system, are trained to fill the gaps at one end caused by death or retirement at the other.

It is important to see that 'society' is a mental construct or model of a web of interrelated social networks intermeshed in a highly complex manner and expanding over geographical and language barriers to cover the whole globe. These networks are 'bunched' into the subsystems of 'society' for purposes of analysis, and boundaries between 'total' social structures are delineated: political systems are perhaps the most fundamental differentiating variable between modern societies. Hence, even if an individual is socialized within British society and is trained to fill a role in the adaptive subsystem or economy, if he in fact emigrates and gives up his British political status to become a South African national, then it is likely he will acquire the political assumptions of that society. Many British nationals may have, for example, kinship links with South Africans without necessarily accepting their political goals or ideals. The importance of the polity can be challenged by the economy (as when it was said by some that 'the city' challenged the Wilson administration in the Autumn of 1964) or the economy can be fused with the polity (as in centrally planned economics), but rule is always rule of the few, whether or not it is accepted by the rest of society. The function of government is to maintain rules, with physical force as an absolute sanction. This does not mean that order implies consensus: certain aspects of the situation may be 'dysfunctional' and may conflict with the goal attainment patterns of some groups within the society but may be 'functional' for others. Similarly the *manifest* function, intended and recognized by the participants in the system, may have a *latent* function producing consequences neither intended nor recognized. As Merton points out in his classic discussion of the subject:

> Through the systematic application of the concept of latent function, therefore, *apparently* irrational behaviour may *at times* be found to be positively functional for the group. Operating with the concept of latent function, we are not too quick to conclude that if an activity of a group does not achieve its nominal purpose, then its persistence can be described only as an instance of 'inertia', 'survival' or 'manipulation by powerful sub-groups in society' (p. 65). . . . Findings concerning latent functions represent a greater increment in knowledge than findings concerning manifest functions. They represent, also, greater departures from 'common-sense' knowledge about social life. (Merton, 1957, p. 68.)

The comparative analysis of new states has helped to drive sociologists back to the larger problems which bothered a generation of scholars at the end of the nineteenth century. However revealing ad hoc explanations may be, they may have little potential for any logical extension and thus the growth of theoretical understanding. Shils sees comparative empirical analysis 'freed from its evolutionary encrustation, and brought into a dynamic conception of social systems' (Shils, 1963, p. 20) as being a more relevant and hopeful approach. He lays stress on the relationship between what he calls 'the centre' and 'the periphery' of society – 'consensus is the key phenomenon of macrosociology . . . How does this institution or practice or belief function in the articulation of the society, in attaching or detaching or fixing each sector in its relationship to the central institutional and value systems of the society?' (pp. 23–24). It is to this consideration of points of tenson and the introduction of change which we now turn.

Parsons has been criticized for being more concerned with what holds societies together than with what drives them on. David Lockwood (1956) has argued that the crucial issue of differential access to scarce resources has been neglected by Parsons and that by stressing the normative aspects he has diverted attention from the *substratum* of social action, especially as it conditions interests which are productive of social conflict and instability. This argument was picked up by Dahrendorf in 1958 who called for a model which would make understandable the structural origin of social conflict. Geographers in particular are likely to have much sympathy with Gouldner's highly apposite criticism of Parsons and Merton. He calls the exclusion of the physical environment by Parsons a form of 'academic monasticism' whereby men are cleansed of their basic passions for sex, food and material possessions by theoretical purification (Gouldner, 1959). The potency of ecological forces has been clearly demonstrated by anthropologists (e.g. Evans-Pritchard 1940) and by Wittfogel in his analysis of hydraulic society. Parsons does not own up to the likelihood of his model of a social system having an unequal capacity to account for variance in social behaviour in different situations. Gouldner goes on to make the crucial point that the consideration of systems in terms of interdependence and equilibrium obscures the fact that they can vary in degree and, further, they may vary differentially. Not all system parts have an equally deep involvement in the resolution of the tensions of the system, or in the mobilization of defences against these. He draws a distinction between functional reciprocity and functional autonomy: those parts with least functional autonomy may be unable to survive separation from a social system and hence may be more likely to be concerned with its conservation than those with greater functional autonomy. Thus it is that the polity may undergo frequent changes in certain Latin American republics without there being any real change in the basic social order, largely because of the polity's functional autonomy. (This example shows that functional autonomy may also militate against charge – a demonstration of Shils' point about central and peripheral subsystems).

Merton, curiously enough, made this point earlier when he said that 'It is not enough to refer to the 'institutions' as though they were all uniformly supported by all groups and strata in the society. Unless systematic consideration is given to the *degree* of support of particular 'institutions' by specific groups we shall overlook the important place of power in society' (Merton, 1957, p. 122). A similar sort of notion can be found in Parsons' work, as we have already noted, but it is not elaborated in any sort of satisfactory way. A valuable statement, which goes a long way to make the functionalist model more flexible, and hence able to accommodate social change, is provided by David Lockwood in his recent paper 'Social Integration and System Integration' (1964). In this he defines social change as 'a transformation of the core institutional order of a society such that we can speak of a change in a type of society' (p. 244). The distinction is made between social integration of the actors as opposed to the integration of the parts of a social system. Normative functionalism provides a way of dealing with both aspects within the same conceptual framework. Lockwood powerfully refutes the conflict theorists, who wish to replace the functionalists' concepts of 'norm', 'consensus' and 'order' with the more fashionable concepts of 'power', 'alienation' and 'conflict', arguing that this is an unreal dichotomy. Conflict is the better grasped within a normative functionalist framework: there is no way of seeing how some conflict is associated with change and other not, except in relation to such a framework. Lockwood goes on to argue that one source of tension and possible change in a social system is 'lack of fit' between its core institutional order and its material substructure. As far as I understand it, he is doing much to integrate crude marxian theory with crude normative functionalism. The 'dominant' or 'core' institutional order may vary from one type of society to another and it is on this that analysis should be focused. Emphasis on the moral aspects of *social* integration has led to neglect of *system* integration by both functional and conflict theorists. We cannot accept that 'The integration of a set of common value patterns with the internalized need-disposition structure of the constituent personalities is the core phenomenon of the dynamics of social systems' (Parsons, 1951, p. 42). We must rather follow Gouldner, Lockwood, Merton and Shils to find a dynamic neo-functionalist model without the Parsonian sweep but with much greater analytical value.

Societies do not change simply through the impact of one factor – whether that factor be coal, geographical position, a decline in the death rate or the growth of an entrepreneurial class. Economic concepts have, perhaps, been accepted too readily by geographers, partly because economists have put forward *theories* of economic development, which are often highly persuasive to the non-economist, and partly because the economist uses quantified indices which are a great solace in a field lacking basic empirical date. The sociologist, whilst not wishing to substitute his 'factors' for those of sister disciplines, nevertheless 'sees' society in terms of a system in which the key points of strain can be isolated. Hence issues such as the flexibility of the system of stratification of that society or the social position of women, may be of crucial importance when considering the goals of a society in relation to its resources. It is most important to remember that 'the goal of the economy is not simply the production of income for the utility of an aggregate of individuals. It is the maximization of production relative to the whole complex of institutionalized value-systems and functions of the society and its subsystems' (Parsons and Smelser, 1956, p. 22). As the goal of the economy is defined by socially structured goals then it is inappropriate to refer to the measurability of utility among individuals. 'Utility, then, is the *economic value* of physical, social or cultural objects in accord with their *significance as facilities* for solving the adaptive problems of social systems' (loc cit.). *The 'spatial constraint' and the endowment of 'natural' resources must be evaluated in the context of a society's adaptive and goal attainment functions.*

Geographers, who have found the models of Myrdal or Rostow useful, should also find *Economy and Society* by Parsons and Smelser helpful in integrating a sociological model into their work. The publication of two admirably concise and reasonably-priced introductory outlines on *Social Change* and *The Sociology of Economic Life,* by Moore and Smelser respectively, should do much to spread these ideas into sixth-form teaching in schools. Indeed, if these books were widely used there would be less need for this chapter in its present form. The emphasis on Parsonian model-building here is felt to be justified because of its very great importance generally and (as far as I know) total neglect among geographers. 'Until Parsons, only economics among all the social sciences could be said to have a rational foundation for its theoretical formulations . . . Parsons has opened the way for other social disciplines to acquire distinctive rationalities of the same type' (Morse, 1961, p. 142).

There are of course a whole range of theories which purport to explain the sources and patterns of change. Herbert Spencer and Auguste Comte in the nineteenth century both believed in the unilinear development of society: in the twentieth century Spengler and Toynbee were less optimistic and saw civilizations rising and falling in waves or cycles. The Marxist theory of society is perhaps the most powerful, if not the most accurate, since it has clearly been able to influence history, even though Marx might be surprised to find his prophecies having more application in, say, Cuba than in England. If Marx believed that revolutions are brought about by a gradual decline from bad to worse then this hardly squares with a comparative analysis of past revolutions. Davies has recently presented a 'Theory of Revolution' (Davies, 1962) in which he argues that 'Revolutions are most likely to occur when a prolonged period of objective economic and social development is followed by a short period of sharp reversal' (p. 6). He goes on to analyse the Dorr's Rebellion of 1842, the Russian Revolution of 1917 and the Egyptian Revolution of 1952. 'Far from making people into revolutionaries, enduring poverty makes for concern with one's solitary self or solitary family at best and resignation or mute despair at worst. When it is a choice between losing their chains or their lives, people will mostly choose to keep their chains, a fact which Marx seems to have overlooked' (p. 9). Davies' model can be diagrammatically represented in Figure 7.1.

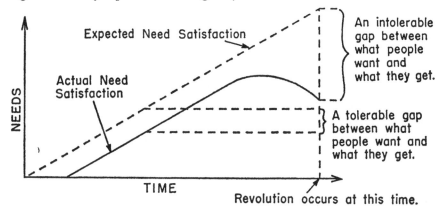

7.1 The relationship between relative deprivation and revolution (*After Davies, 1962*).

Moore makes a valuable distinction between 'mere *sequences of small actions*, that in sum essentially *comprise the pattern*, the system, and *changes in the system* itself, in the magnitude or the boundaries, in the prescriptions for action, in the relation of a particular system to its environment' (Moore, 1963, p. 6). Neither rebellions, nor indeed revolutions, necessarily lead to changes in the system: that is part of the tragedy of revolutions, but it also illustrates Gouldner's concept of functional autonomy. Moore underlines the point that we have already made – namely that an equilibrium model of society 'either forecloses questions about the sources of change, or if discordant internal elements are brought into the analysis, the theoretical model will predict one direction of change, and one only – change that restores the system to a steady state' (p. 10). If, on the other hand, society is seen as a tension-management system, where both order and change are problematical but also 'normal', then there is no need to expect that change to accommodate tension will necessarily remove tension, but will rather shift it to other points in the system. The crucial issue, which Moore very neatly poses, is thus to distinguish between changes in the *system* and changes which are inherent in the way the system operates. It is Moore's contention that one of the main change-producing strains which impinges on a tension-management social system is provided by the non-human environment.

It is somewhat ironic that sociologists should now be turning to the broad issues that geographers may feel they outgrew at the end of the nineteenth century. At the sixtieth annual meeting of the American Sociological Association, held in 1965, it is significant that the first plenary session was devoted to a re-evaluation of Karl Marx (in which Parsons argued that Marx was not sufficiently alert to other forms of conflict in society than that based on class) and the second session was devoted to 'Civilizations and their changes'. There seems no doubt that these macrosociological problems of social and cultural change, based on the models discussed in this chapter, will continue to exercise the minds of some of the most distinguished sociologists in the world. It would be unfortunate if geographers, preoccupied with models of man at a sub-social, materialist or mechanistic level, withdrew from their traditional interest in this field. So much of the discussion by geographers seems to take place in inaugural lectures, when the determinism – possibilism skeleton is taken out of the cupboard for its ritual dusting, that there is a danger that younger geographers will neglect the field for fear of getting tarred with the brush of their elders. Yet if one turns to the collection of readings on *Social Change* edited by the Etzionis, for example, the range and fertility of the *ideas* of contemporary sociologists is impressive. There is clearly a danger that over-concern with techniques may be at the expense of conceptual model building in geography. Certainly work by sociologists over the past ten years – particularly by Parsons, Smelser and Moore and the others mentioned above – is moving towards some general theory of change for whole societies. Smelser's recent essay '*Toward a Theory of Modernization*' is worth a long and careful look by geographers. He attempts to construct an ideal type, in the Weberian sense, to analyse the relationship between economic growth and social structure: in particular he builds up a *differentiation* model to clarify the way major social functions gain structural independence. For every social function there is a distinct set of structural conditions under which it is optimally served and this is the link between modernization and differentiation. Historical geographers would no doubt find Smelser's study of the Lancashire cotton industry which 'twisted, spiralled, reversed, and creaked as it accumulated the elements which carried it nearer to Weber's conditions of extensive role differentiation' (Smelser, 1959, p. 101) a particularly relevant introduction to his model.

In the light of Smelser's ideas it is useful to consider Epstein's study of *Economic Development and Social Change in South India*, based on the two villages of Wangala and Dalena. She shows convincingly that economic development does not necessarily lead to economic change. In the case of Wangala the introduction of irrigation changed the social system very slightly, as the traditional farming economy was simply strengthened. Only where the new economic system was incompatible with features of traditional economic organization did change in economic roles and relations occur. This was the case at Dalena, where the whole social system was changed when the village had to provide secondary and tertiary services for the neighbouring villages, where the land had been irrigated. The 'backwash' effects of the change in the economic system led to changes in political and ritual roles and relations and in the principles of social organization. Such empirical studies as this in different social and physical environments are urgently needed if the conceptual models of social change and modernization are to be refined and developed.

SOCIOLOGICAL MODELS AND URBAN GEOGRAPHY

It may be that geographers are more willing to accept quantitative techniques from elsewhere than conceptual models, and yet in certain areas the very lack of a conceptual base seriously impedes research. Much of the apparent precision of recent studies in locational analysis is based on inadequate operational definitions of populations. As Haggett (1965, p. 189) admits, 'The problem of standardizing definitions of cities has not been solved'. Furthermore, Berry, when discussing the disappointing results of research by geographers on the residential pattern in cities, concludes that 'geographers must take second place to urban sociologists in studies of the residential patterning of cities' (in Hauser and Schnore, 1965, p. 417).

Now why should this be so? Why must the geographer admit defeat both in spatial delimitation and in the understanding or explanation of a spatial pattern? The answer, surely, must be that he is working with a wrong model and is approaching the problem with the wrong assumptions. Norton Ginsburg is an urban geographer who has moved some way towards a more sociologically informed approach to urbanization, no doubt stimulated by the seminal atmosphere of the admirable multi-disciplinary tradition of the University of Chicago. Ginsburg outlines his problem thus: 'What kinds of cities can be expected to evolve in different societies as these societies make their decisions to select, adopt, and modify those elements that characterize Western city-building, functions, and structure?' (in Hauser and Schnore, 1965, p. 319) and he goes on to describe Japan as a highly urbanized country, not in demographic terms – the proportion of people living in sizable towns – but in more *sociological* terms of the impact of the city on the nation's life and 'the awareness among rural dwellers of the existence and nature of an urban way of life'.

Ginsburg appears to be right in making a distinction between ways of life, for how else is one to distinguish between, say, an Indian city, parts of which may be little more than villages joined together with a 'rural' pattern of life within them, and the situation in a Japanese city, the product of a different society with a strong tradition of urbanism? There is clearly no reason why geographers should not concern themselves with sociological problems – so long as they remember that differences which are internal to a social system cannot be explained only by reference to forces external to it. Those who see 'the necessities of economic expansion' as a sort of global 'prime mover' in stimulating 'urbanization' and an increase in the 'scale of society' have been criticized by Sjoberg for failing to explicate their theories (in Hauser and Schnore, 1965, pp. 168–177) Berry and Ginsburg have been more honest in rejecting the adequacy of non-sociologically-based frameworks.

Work by Duncan and Schnore (1959), influenced by Durkheim's *Division of Labour in Society* suggests a model with four basic components – environment, population, social organization and technology – which are functionally interrelated. Sjoberg accuses them of materialism and finds the concept of social organization 'particularly spongy'. Perhaps the most sensible statement on whether settlement size and other ecological concepts are useful for explaining ways of life has been provided in a succinct and satisfying essay by Gans (in Rose, 1962, pp. 625–648) and he deserves to be cited at length:

> Ecological explanations of social life are most applicable if the subjects under study lack the ability to *make choices*, be they plants, animals, or human beings. Thus, if there is a housing shortage, people will live almost anywhere, and under extreme conditions of no choice, as in a disaster, married and single, old and young, middle and working class, stable and transient will be found side by side in whatever accommodations are available. At that time, their ways of life represent an almost direct adaptation to the environment. If the supply of housing and of neighbourhoods is such that alternatives are available, however, people will make choices, and if the housing market is responsive, they can even make and satisfy explicit *demands*.
>
> Choices and demands do not develop independently or at random; they are functions of the roles people play in the social system. These can best be understood in terms of the *characteristics* of the people involved; that is characteristics can be used as indices to choices and demands made in the roles that constitute ways of life. Although many characteristics affect the choices and demands people make with respect to housing and neighbourhoods, the most important ones seem to be *class* – in all its economic, social, and cultural ramifications – and *life-cycle stage*. If people have an opportunity to choose, these two characteristics will go far in explaining the kinds of housing and neighbourhoods they will occupy and the ways of life they will try to establish within them (pp. 639–640).
>
> Characteristics do not explain the causes of behaviour; rather, they are clues to socially created and culturally defined roles, choices and demands. A causal analysis must trace them back to the larger social, economic, and political systems which determine the situations in which roles are played and the cultural content of choices and demands, as well as the opportunities for their achievement (p. 641).

Gans goes on to conclude that if ways of life do not coincide with settlement type but are rather functions of class and life-cycle stage, then a sociological definition of the city cannot be formulated.

With these ideas in mind it is possible to probe the implicit model in my study of the rural/urban fringe of the London Metropolitan Region (Pahl, 1965). This was an attempt to study the 'urban' influences on a 'rural' area and illustrates very well the need to abandon traditional geographical concepts for a sociological approach in such a problem area. Indeed I started with a naïve assumption that it would be possible to use a 'spatial' model so that 'accessibility' to urban functions could be taken as a key variable. It soon became clear that to group the whole population together, as so many social atoms showing certain mathematical regularities, was a meaningless exercise. The styles of life of two broad sub-groups of the population, empirically determined, were so very different that a non-spatial conceptual model was forced upon me. For members of the working class, who commute to work out of the area, space is a 'constraint' which has to be overcome and the cost of doing so is an added economic burden on the family. Not all the working class *choose* to live in a rural area but they are 'forced' to do so by a society which allocates council houses by area residence and not place of present employment.

By contrast, for a section of the middle class, space, far from being a constraint, is valued as an amenity, which should be preserved. Space may become a symbol of a certain style of life to which such middle-class people aspire and the economic burden of crossing it, in order to reach urban employment and amenities, is an accepted concomitant of their way of life. Such people *choose* to live in an area where working class people may be *forced* to live and the residential pattern is a resolution of the two forces. Clearly this polar ideal type can be made more complex by adding further groups – for example, those who aspire to the middle-class life style but whose position in the larger socio-economic system makes it difficult for them to maintain it. A further group may be obliged to move into the rural-urban fringe area simply because the price of land nearer towns makes home ownership there impossible. Yet another group, at the top of the social hierarchy, may feel that the power and influence which they wish to achieve is more easily had by living in the centre of cities and then their country house becomes of secondary importance and is only used at week-ends or at certain times of the year.

I tried to show in my study that any attempt to understand the striking expansion of population in the Outer Metropolitan Region, outside the officially designated Urban Districts, cannot be achieved with simple materialistic models, be they called ecological, spatial, geographic, demographic or whatever. Yet this does not mean that a consideration of people's values leads to a confusion without order. People's choices reflect values that are shared. Other people find themselves in a similar position in the socio-technical system and, with more limited house-buying potential, are forced to certain areas. Yet other people occupy an even more restricted position in the social system so that they do not even have the choice of owning a home; and so on. The residential pattern is a reflection of the functioning of the social system. In Britain we have a conventional socio-economic stratification system related to the unequal distribution of social and economic resources, which is supported by the central value system of our society. In other societies with a different social system – say India – the residential pattern of the population will also be different. Yet in both societies the poorest will have no choice, so that their position can be explained with a materialistic model and they will be more amenable to study with quantitative techniques based on 'economic man'. At the other extreme the very rich have the greatest choice in where they shall live and what life-styles they will pursue (cf. Pahl, 1966).

If the geographer is not to admit complete defeat in accounting for the residential pattern he must try to understand the sociologists' model of social stratification for the appropriate society, together with a knowledge of the values, in particular the aspirations for certain life styles of major sub-groups of the population; this, as Gans pointed out in the quotation above, will lead the geographer to the larger social, economic and political systems. Certainly geographers must be sure to make a distinction in their quantitative analyses between least-cost locations, which involve no choice, and other locations based on other values. Failure to do this may lead the most exciting breakthrough of this century to lose its momentum in the materialistic determinism of a 'social physics', trapping the model makers in a strait-jacket of their own invention. It is vital that interdisciplinary co-operation and criticism continues: a fitting conclusion is provided by the following comment – the result of co-operation between an economist and a social anthropologist:

> Whether particular models and the development of particular lines of research are worthwhile or not must in the end be a matter of personal judgment and faith. All that we can plead for is that those who engage in this activity should not arrogantly assume that their system is necessarily the best or that their system will necessarily be relevant to the explanation of reality, or that a model which appears to explain development in one country will have universal application and validity. (Gluckman and Devons, 1964, p. 186.)

REFERENCES

ARON, R. and HOSELITZ, B. F., (Eds.), [1965], *Social Development*, (Paris).

ABRAMS, P., [1963], Notes on the Uses of Ignorance; *Twentieth Century*, 67–77.

BECKER, H. and BARNES, H. E., [1961], *Social Thought From Lore to Science*, (New York).

BENDIX, R., [1962], *Max Weber. An Intellectual Portrait*, (New York).

BESHERS, J. M., [1957], Models and Theory Construction; *American Sociological Review*, 22, 32–38.

BLACK, MAX. (Ed.), [1961], *The Social Theories of Talcott Parsons*, (Englewood Cliffs, N.J.).

BOSKOFF, A., [1964], Functional Analysis as a Source of a Theoretical Repertory and Research Tasks in the Study of Social Change; In Zollschan, G. K. and Hirsch, W., (Eds.), *Explorations in Social Change*, (London), pp. 213–243.

BOTTOMORE, T. B., [1962], *Sociology*, (London).

BRODBECK, M., [1959], *Models, Meaning and Theory*: In Gross, L., (Ed.), *Symposium on Sociological Theory*, 275–403.

COLEMAN, J. S., [1964], *Mathematical Sociology*, (London).

DAHRENDORF, R., [1957], *Class and Class Conflict in Industrial Society*, (London).

DAHRENDORF, R., [1958], Toward a Theory of Social Conflict, *Journal of Conflict Resolution*, 11, 170–183.

DAVIES, J. C., [1962], Toward a Theory of Revolution; *American Sociological Review*, 27, 5–19.

DUNCAN, O. D. and SCHNORE, L. F., [1959], Cultural, Behavioural and Ecological Perspectives in the Study of Social Organization; *American Journal of Sociology*, 65, 132–146.

DURKHEIM, E., [1947], *The Division of Labour in Society*, (trans. G. Simpson), (Glencoe, Illinois), (First published in 1893).

EISENSTADT, S. N., [1963], *The Political Systems of Empires: The Rise and Fall of Historical Bureaucratic Societies*, (London).

EPSTEIN, T. S., [1962], *Economic Development and Social Change in South India*, (Manchester).

ETZIONI, A. and E., (Eds.), [1964], *Social Change: Sources, Patterns and Consequences*, (London).

EVANS-PRITCHARD, E. E., [1940], *The Nuer*, (Oxford).

GERTH, H. H. and MILLS, C. W., [1948], *From Max Weber: Essays in Sociology*, (London).

GLUCKMAN, M. and DEVONS, E., (Eds.), [1964], *Closed Systems and Open Minds*, (Edinburgh).

GOULD, J. and KOLB, W. L., (Eds.), [1964], *A Dictionary of the Social Sciences*, (Paris).

GOULDNER, A., [1959], Reciprocity and Autonomy in Functional Theory; In Gross, L., (Ed.), *Symposium on Sociological Theory*, 241–270.

GOULDNER, A. W. and H. P., [1963], *Modern Sociology*, (London).

GROSS, L. (Ed.), [1959], *Symposium on Sociological Theory*, (Row, Peterson & Co.).

HAGEN, E. E., [1962], *On the Theory of Social Change*, (Homewood, Illinois).

HAGGETT, P. [1965], *Locational Analysis in Human Geography*, (London).

HAUSER, P. M. and SCHNORE, L. F., (Eds.), [1965], *The Study of Urbanization*, (London).

HEMPEL, CARL, [1952], Typological Methods in the Social Sciences; Reprinted in M. Natanson, (Ed.), *Philosophy of the Social Sciences: A Reader*; New York: Random House, 1963, pp. 210–230 (see section on Ideal Types and Theoretical Models, pp. 223–230).

HUGHES, H. S., [1959], *Consciousness and Society: The Reorientation of European Social Thought 1890–1930*, (London).

INKELES, A., [1964], *What is Sociology?*, (Englewood Cliffs, N.J.), (Ch. 3 Models of Society in Sociological Analysis).

JOHNSON, H. M., [1961], *Sociology*, (London).

LA PIERE, R. T., [1965], Social Change, (London), (Ch. 3 Models of Change and Stability).

LOCKWOOD, D., [1956], Some Remarks on 'The Social System'; *British Journal of Sociology* 7 (2).

LOCKWOOD, D., [1964], Social Integration and System Integration; In Zollschan, G. K. and Hirsch, W., (Eds.), *Explorations in Social Change*, (London), 244–257.

MEADOWS, P., [1957], Models, Systems and Science; *American Sociological Review*, 22, 3–9.

MERTON, R. K., [1957], *Social Theory and Social Structure*, (Glencoe, Ill.).

MILLS, C. W., [1959], *The Sociological Imagination*, (New York), (Ch. 2 Grand Theory).

MOORE, E., [1963], *Social Change*, (Englewood Cliffs, N.J.).

MOORE, W. E. and HOSELITZ, B. F., (Eds.), [1963], *Industrialization and Society*, (Paris and the Hague).

MORSE, C., [1961], The Functional Imperatives; In Black, M., (Ed.), *The Social Theories of Talcott Parsons*, (Englewood Cliffs, N.J.), 100–152.

PAHL, R. E., [1965], *Urbs in Rure*; London School of Economics Geographical Paper No. 2.

PAHL, R. E., [1966], The Rural-Urban Continuum; *Sociologia Ruralis*, 6 (3–4), pp. 299–326.

PARSONS, T., [1937], *The Structure of Social Action*, (New York).

PARSONS, T., [1951], *The Social System*, (London).

PARSONS, T. and SHILS, E. A., [1951], *Toward a General Theory of Action*, (Harper Torchbook Edn., New York 1962).

PARSONS, T., [1954], Present Position and Prospects of Systematic Theory in Sociology (1945); In *Essays in Sociological Theory*, (Glencoe, Ill.), 212–237.

PARSONS, T. and SMELSER, N. J., [1956], *Economy and Society*, (London).

PARSONS, T., [1960], The Principle Structures of Community; In *Structure and Process in Modern Societies*, (Glencoe, Ill.), 250–279.

PARSONS, T., [1961], Some Considerations on the Theory of Social Change; *Rural Sociology*, 3, 219–239.

RADCLIFFE-BROWN, A. R., [1952], *Structure and Function in Primitive Society*, (London).

PONSIOEN, J. A., [1962], *The Analysis of Social Change Reconsidered*, ('S – Gravenhage, Netherlands).

ROSE, A. M., (Ed.), (1962), *Human Behaviour and Social Processes*, (London), (Chs. 33 and 34).

REX, J., [1961], *Key Problems in Sociological Theory*, (London).

SHILS, E. A., [1963], On the Comparative Study of the New States; In Geertz, C., (Ed.), *Old Societies and New States*, (London).

SMELSER, N. J., [1959], *Social Change in the Industrial Revolution*, (London).

SMELSER, N. J., [1963], *The Sociology of Economic Life*, (Englewood Cliffs, N.J.).

SMELSER, N. J., [1964], Toward a Theory of Modernization; In Etzioni, A. and E., (Eds.), *Social Change: Sources, Patterns, and Consequences*, (London), 258–274.

STEIN, M. and VIDICH, A., (Eds.), [1963], *Sociology on Trial*, (Englewood Cliffs, N.J.), (esp. essays by Gouldner and Foss).

WEBER, M., [1947], *The Theory of Social and Economic Organisation*, (New York).

WITTFOGEL, K. A., [1957], *Oriental Despotism*, (New Haven and London).

ZOLLSCHAN, G. K. and HIRSCH, W., (Eds.), [1964], *Explorations in Social Change*, (London).

CHAPTER EIGHT

Models of Economic Development

D. E. KEEBLE

GEOGRAPHY MODELS AND ECONOMIC DEVELOPMENT

Any examination of the professional geographical literature of recent years reveals an apparent and remarkable lack of interest among geographers in the study of the phenomenon of 'economic development' (Ginsburg, 1960, p. ix; Mountjoy, 1963, p. 13; Steel, 1964, p. 13; Lacoste, 1962, p. 248). For example, of the 251 major articles (excluding editorials and reviews) published between 1955 and 1964 inclusive in what is probably the most relevant professional geographical journal, *Economic Geography*, only ten were explicitly concerned in whole or part with problems of economic development. With the more general journal of the *Annals of the Association of American Geographers*, the percentage falls still further, to 2·5 per cent (i.e. six articles out of 242). This state of affairs is remarkable in view both of traditional geographical concern with countries now categorized as 'underdeveloped', and of the enormous surge of interest in problems of economic development which has occurred in other, often fairly closely allied, disciplines (e.g. history, sociology, politics and economics) since the Second World War (Goldsmith, 1959, p. 25; Meier and Baldwin, 1957, p. 1; Pen, 1965, p. 190; Gerschenkron, 1962, pp. 5–6; Meynaud, 1963, pp. 9–10).

The background to this apparent disinterest is undoubtedly complex. But in addition to 'the extreme separatism of geographers as a group' (Haggett, 1965A, p. 101; see also Chorley and Haggett, 1965, p. 375; Ackerman, 1963, pp. 431–432; Chisholm, 1966, pp. 1–2), geography's traditional preoccupation with the individuality and uniqueness of different countries and areas – i.e. the 'idiographic' approach – rather than with their general similarities – i.e. the 'nomothetic' approach (Hartshorne, 1939, pp. 378–384; Ackerman, 1958, pp. 13–16; Bunge, 1962, pp. 7–13) must surely have played some part. For decades, geographers have concerned themselves primarily with the description and analysis of those unique combinations of spatially-associated phenomena which are found in particular, individual, areas and countries (Hartshorne, 1959, pp. 146–149). The categorization in the post

Second World War period of many of these areas as 'underdeveloped' – i.e. part of a general group of areas, dominated by common features and problems – has failed to influence this attitude, a failure evident in the disinterest in economic development problems noted above.

But the effects of this idiographic approach go farther still. For even those geographical studies which focus explicitly on the problems and nature of economic development bear its imprint. Such studies may be divided into the four groups listed in Table 8.1, which also indicates the distribution among these groups of the 16 *Annals* and *Economic Geography* articles mentioned earlier.

TABLE 8.1

'Geographical' Articles on Economic Development – A Classification

Group	Main Focus of Article's Interest	Number of Articles
A	Relationship between Physical Environment (especially Natural Resources) and Economic Development	4
B	Classification of Areas in terms of Indexes of Economic Development	3
C	Unique Characteristics of an Individual Area, with peripheral reference to its Economic Development	6
D	Other	3
	Total	16

Although the sample presented above is perforce a small one, examination of other geographical studies relating explicitly to economic development strongly supports the fourfold grouping suggested. Into Group A, which already includes Ginsburg's *Natural Resources and Economic Development* (1957), and Tosi and Voertman's analysis (1964) of relationships between physical environment and economic development in the tropics, fall such studies as those of James (1951), Keller (1953), Stamp (1953 and 1963), Gribaudi (1965) and Fordham (1965). Group B, including Fryer's classificatory article (1958) in *Economic Geography*, is represented on the wider scale by the opening pages of his *World Economic Development* (1965, pp. 3–24), by Ginsburg's stimulating *Atlas of Economic Development* (1961), and by many of the contributions to the latter's pioneering *Essays on Geography and Economic Development* (1960), notably those by Hartshorne, Wagner, Guyol, Berry, Gosling and Rodgers. Amongst many further examples of Group C may be singled out the various case-studies included in

Hance's *African Economic Development* (1958), together with Ooi Jin-Bee's study of rural development in Malaya (1959), Dwyer's excellent analysis (1965) of Hong Kong, and Green and Fair's work (1962) on economic growth in Southern Africa. Only a very few geographical studies, such as Mountjoy's (1963), cannot be allocated to the first three groups, and must therefore be added to Group D.

Of the four groups, Group C, with the most articles in the two journals examined, is clearly closely identified with the idiographic approach. The accent throughout these articles is on the individual analysis of the particular areas chosen. The same approach is evident, however, in studies in Group A, since apart from Ginsburg's article (1957), most of these work explicitly or implicitly from the viewpoint that 'the specific variations imposed by the conditions of the total environment are unique', and that in analysing the effect of the physical environment upon economic development 'general concepts of wide applicability may be less important than the careful analysis of unique situations' (James, 1951, p. 230). Even studies in Group D appear to emphasize this idiographic approach at times. For example, Mountjoy devotes two chapters to exemplifying his comment that 'every nation runs an individual course in the cross-country development race' (Mountjoy, 1963, p. 157): while Orchard (1960) is undoubtedly as interested in East and South Asia as a unique area as in utilizing it as 'a laboratory for the economic geographer' (Orchard, 1960, p. 215). Only in Group B are the majority of studies concerned explicitly with a search for underlying similarities in patterns of economic development. The nomothetic approach of such classificatory work is exemplified by Ginsburg's study (1960), which concludes that geography's contribution to the study of economic development 'will be greatest through the careful examination and analysis of reality so as to test, appraise, and modify generalizations, rather than through the idiographic study of presumably isolated events' (Ginsburg, 1960, p. xx).

The predominantly idiographic approach of what little geographical analysis of economic development has been undertaken stands in marked contrast to the approach adopted in other disciplines. Only amongst historians is the study of individual, unique, cases of economic development of major disciplinary importance (Hicks, 1953; Hoselitz, 1959); and even here, the search for more general similarities and concepts appears to have intensified in recent years (Hoselitz, 1955A and 1959; Supple, 1963, pp. 7–8; Conrad and Meyer, 1965, pp. 3–28; Rostow, 1960, p. 1). In other disciplines, particularly the most significant one in this context, economics, the nomothetic approach predominates. Indeed, Goldsmith (1959, p. 27) for one claims that idiographic attitudes have no place whatever in economic analysis as applied to problems of economic growth. This major contrast between geography and economics clearly reflects the different historical origins of the two disciplines (McNee, 1959; Chisholm, 1966, pp. 4–25), the former's largely 'empirical and descriptive' character stemming from nineteenth-century Darwinian thinking, the latter's 'more abstract' character as 'a deductive system of logic' being derived chiefly from eighteenth-century rational thinking (McNee, 1959, p. 191).

The nomothetic approach of economics finds its greatest expression in the construction of economic models. An economic model may be defined as 'an organized set of relationships that describe the functioning of an economic entity . . . under a set of simplifying assumptions' (United Nations, 1961, p. 7). By selecting those aspects of economic reality which are deemed particularly significant, and concentrating on the relationships between these few aspects, economists have developed models of economic activity which are 'of great value in interpreting, if not predicting, economic behaviour in the real world' (McNee, 1959, p. 191). Model-building of this kind has in recent years become a key element in economic analysis (Pen, 1965, p. 65; Orcutt, 1960, p. 897). In particular, it has been very rapidly applied to the study of economic development, both conceptually and for planning purposes. This is not to say that economists have not followed other approaches. Even studies of the influence of natural resources upon development (Spengler, 1961; Clawson, 1964) and of methods of classifying countries according to level of development (El-Kammash, 1963) have been included in the vast flood of economic literature on economic development in the last decade. But the building, testing and application of growth models has undoubtedly formed a very important, if not dominant, part of economics' contribution to the study of economic development.

Until recently, one major defect of this model-building activity was its lack of concern with the spatial changes inherent in economic growth. Since the Second World War, however, economics has at last come to realize the significance of this omission, and begun 'balancing its spaceless models with others including the spatial variable' (McNee, 1959, p. 198; see also Chisholm, 1966, p. 2). Despite certain problems (Meyer, 1963, p. 41; Paauw, 1961, p. 180), this process has proceeded so rapidly that a general survey of model-building attempts to analyse the spatial aspects of economic growth is fully justified. Such a survey is perforce concerned primarily with the work of economists who, unfettered by an idiographic tradition, have at last moved to fill the wide intellectual void left open by geographers.

Figure 8.1 portrays diagrammatically a simple typology of economic growth models, organized on the basis of both spatial content and scale coverage. The breakdown of the latter into Supra-National, National and Sub-National units, quite apart from its similarity to scale classifications adopted by other analyses of similar topics (Isard and Reiner, 1961, p. 19; Isard and Smolensky, 1963, p. 105; Friedmann, 1963, pp. 43–44), may be justified in two ways. Firstly, on practical grounds, most models have been developed specifically in terms of one or other of these scale groups. Secondly, on conceptual grounds, both the degree of internal homogeneity of those factors influencing economic development – government economic policy, laws, currency, language, financial institutions, communication systems, etc. – if not of the level of economic development itself, and the degree of 'openness' of the economy concerned to external economic stimuli, vary fairly sharply between these groups.

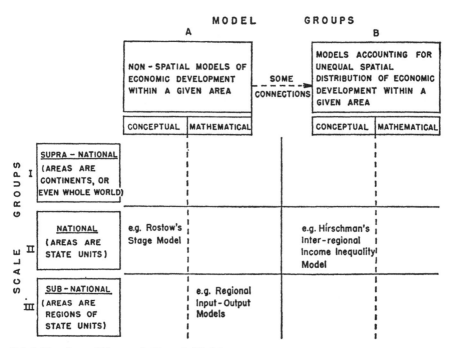

8.1 A Typology of Economic Growth Models.

The key scale group here is undoubtedly the national one. Most economists (e.g. Kuznets, 1951; Robinson, 1960, pp. xiv–xv) appear to agree that 'in analysing economic growth and structure . . . national states are the natural units of comparison' (Goldsmith, 1959, p. 23). State units generally exhibit not only a high level of internal economic homogeneity but also a low degree of openness, in that they are surrounded by considerable barriers to the free flow of commodities — labour, capital, goods, ideas – vital to the process of economic growth (Kuznets, 1951, pp. 29–33; Robinson, 1960, p. xiv). Interestingly enough, both these criteria appear to vary to some extent with the overall level of development of the state unit concerned (Williamson, 1965; Berry, 1960), although this cannot be pursued here. At scales greater than the national unit, however – i.e. continental and world scales, both of which fall into the Supra-National group – internal homogeneity and degree of openness become far less. Indeed, at the world scale, the economic system under consideration is of course completely closed (Tiebout, 1956, p. 161). Conversely, at the Sub-National scale, the economic system becomes much more open, often with virtually complete mobility of commodities across its borders (Harris, 1954, p. 369; Sickle, 1954, p. 382) – while the level of internal homogeneity may also rise. The last point is of course particularly true of those regions which have been delimited in terms of homogeneous developmental characteristics – an exercise in which geographers have hitherto played very little part, despite the discipline's traditional preoccupation with 'regional geography' (see McLoughlin, 1966). For the above reasons, therefore, it appears more realistic to organize the scale breakdown in terms of national units (and subdivisions, or aggregations, of these) than in terms of some absolute scale index, such as that proposed by Haggett, Chorley and Stoddart (1965). The variability in absolute size of state units, and its influence upon levels of economic development, will however be referred to later (see p. 254).

Although it is clear that geographers will primarily be interested in Group B models, certain aspects of those in Group A may also be of value in geographical study. These latter models will be discussed first.

NON-SPATIAL MODELS OF ECONOMIC DEVELOPMENT

The great majority of the interpretations and models of economic growth produced by economists have been non-spatial, a situation arising from the early 'elimination of the spatial variable from theoretical economics' (McNee, 1959, p. 192). None the less, it seems useful to consider some of them in order to illustrate the ways in which economic models have been constructed, to indicate techniques which could be adapted to the building of spatial models, and to demonstrate the possible use of growth models for classifying and comparing different economies. In that an economic model is basically only 'a simplified description of reality' which stresses 'crucial variables' at the expense of the 'myriad of variables that are of secondary importance' (Borts and Stein, 1964, p. 48), apparently dissimilar intellectual constructs may none the less rightly be termed models. For example, a verbal analysis of the generalized pattern of economic change which is thought to have occurred in advanced economies is as much a model as any set of algebraic formulae used under mathematically-specified assumptions to calculate changing values of economic parameters (Enke, 1964, p. 189; Pen, 1965, p. 65). The former type of construct will be referred to here as a conceptual/historical model, the latter as a mathematical model.

National-scale non-spatial models

(1) *Conceptual/historical models*. Although acknowledging the difficulty of reducing to some common denominator the economic history of such widely differing nations as Britain, the United States and Denmark, Hoselitz (1959) has stressed the need for economic growth models to be based on the history of already developed countries. Such models should single out 'from the unique economic history of each country . . . some of the variables that seem to have a crucial impact' (Hoselitz, 1959, p. 146), so that the process of growth in these different economies may be more easily compared. At the same time, however, Paauw (1961) has emphasized that the building of historical growth models on the basis of 'armchair theorizing or almost intuitive observation' has 'tended to outstrip empirical verification'. Such models must therefore also be 'formulated as meaningful propositions (i.e. propositions which are verifiable, or may be refuted, if only under ideal conditions)' (Paauw, 1961, p. 180). Finally, Enke (1964, p. 189) has pointed out that any model based on historical data 'must be much more than a description' of growth regularities. It should rather develop 'its own inner logic' (Enke, 1964, p. 190), by showing how changes during one period of growth are related to those occurring during preceding and subsequent ones.

These criteria are very useful in assessing proposed conceptual/historical models. The earliest such were the 'stage theories' of certain late nineteenth-century German economists, such as List, Hildebrand, Bucher and Smoller. These models organized the economic history of the then advanced countries into stages distinguished from each other by a variety of criteria – by character of exchange system in Hildebrand's model (barter; money; credit), by dominant occupations in List's (savage; pastoral; agricultural; agricultural and manufacturing; agricultural, manufacturing and commercial) (Enke, 1964, pp. 191–194). However, these early stage models were in fact little more than descriptive classifications of different types of economic organization supposed to follow each other in time (Goldsmith, 1959, p. 25): and when tested against the real world, their extremely low level of applicability rendered them virtually useless as analytical tools (Hoselitz, 1960, pp. 193–238; Meier and Baldwin, 1957, pp. 143–147; Gras, 1930).

The attempt to develop a more sophisticated and meaningful stage model of economic growth has only been made in quite recent years. Since its initial formulation (1955, Chap. 7, and 1956), extension (1959) and final elaboration (1960), Rostow's generalization of 'the sweep of modern history' as 'a set of stages-of-growth' (Rostow, 1960, p. 1) has attracted remarkable attention. Accepted with alacrity both by economists concerned with underdeveloped countries (Haq, 1963, p. 13; Das-Gupta, 1965, p. 56; Enke, 1964, p. 201; Meier, 1964, p. 3; Brandenburg, 1964, p. 96) and by the intelligent public (Ohlin, 1961, p. 648; Magee, 1965, p. 76), it has none the less provoked strong criticism from economists in developed countries. Its five stages are portrayed diagrammatically in Figure 8.2. Rostow's own chart applying the model to particular countries is reproduced in Figure 8.3.

The model's base-level, as it were, is the traditional society, characterized by limited technology, pre-Newtonian attitudes to science and the physical

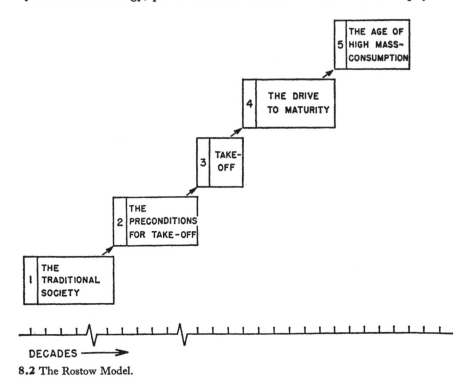

DECADES ⟶

8.2 The Rostow Model.

world, and a static and hierarchical social structure. As primarily exogenous influences stimulate the beginnings of a rise in the rate of productive investment, the installation of 'social overhead capital' (roads, railways, etc.), and the evolution of a new social/political elite, the preconditions for take-off stage develops, with agriculture and extractive industry playing a key role. The crucial stage, however, is take-off, the 'decade or two' when economy and society 'are transformed in such a way that a steady rate of growth can be, thereafter, regularly sustained' (Rostow, 1960, pp. 8–9). In practical terms, take-off is launched by some initial stimulus, and characterized by 'a rise in the rate of productive investment . . . to over 10 per cent of national income', the 'development of one or more substantial manufacturing sectors, with a high rate of growth', and the emergence of 'a political, social and institutional framework' which encourages growth (Rostow, 1960, p. 39). After take-off follows the drive to maturity, during which the impact of growth is transmitted to all parts of the economy. Finally, with the shift in sectoral leadership to industries such as durable consumer goods, ensues the age of high mass-consumption – although other alternatives, such as the pursuit of international power, or of the welfare state, may be chosen by particular societies instead.

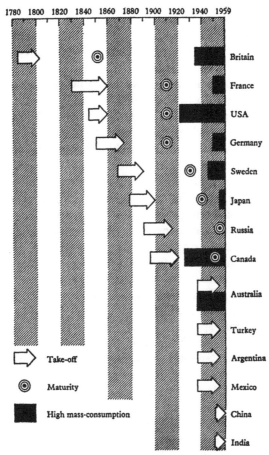

8.3 The Rostow Model Applied to Selected Countries.
(*Source: Rostow, 1960, p. xii.* The chart was originally published in *The Economist*, August 15, 1959, p. 413).

The model's 'provocative suggestions' (Hagen, 1962, p. 522) represent an attempt to 'isolate the strategic factors' in economic growth which is undoubtedly 'more substantial . . . more analytical and related to a wider range of issues' than that of any previous stage model-builder (Meier, 1964, p. 25). Despite this, however, numerous criticisms have been levelled against it – criticisms which appear to operate on three distinct levels. On the first level,

Rostow is criticized for having tried to build a model at all. For example, Cairncross (1961, p. 451) claims that any attempt to explain economic growth in terms of 'one or two all-embracing variables' is impossible: growth patterns vary too greatly between different countries. This criticism is to a large extent refuted by Hagen (1962) and Higgins (1964), who agree that the search for regularities in growth patterns is a very necessary task; and that what really matters is not whether Rostow's model and its stages 'ignore some complexities of reality, but whether they are congruent with reality in respects which make them useful for its analysis' (Hagen, 1962, p. 514). This point will be discussed further below.

The second level of criticism is that which accepts the value of model-building, but criticizes Rostow's model as being a poor example of the species. In other words, it attacks the model *as a model*. At this level, the three criteria listed earlier provide valuable yardsticks for assessment. Does Rostow's model single out crucial variables, define its concepts precisely enough to permit comparison with the real world, and possess an analytical rather than descriptive framework? On the first score, the model comes out well. In stressing, for example, the role of 'leading sectors' (i.e. economic activities which exhibit high growth rates and from which growth is transmitted to other parts of the economy), of agriculture and the provision of social overhead capital during early stages in growth, and of a growth-orientated political and social framework, Rostow does appear to have focused attention on variables judged very important by other economists (Hagen, 1959, p. 135; Hoselitz, 1959, pp. 154-155; Johnston and Mellor, 1961; Meier, 1964, pp. 266-272; Meynaud, 1963). Over the second criterion, however, commentators appear divided. Hagen (1962, p. 515), for example, applauds Rostow for identifying 'specific conditions whose presence or absence can be tested'. Certainly this appears true of one of the most crucial variables in his model, the concept of a sudden sharp rise in rate of productive investment at take-off. However, other observers (Enke, 1964, p. 201; Cairncross, 1961, p. 451; Drummond, 1961, p. 113) strongly disagree. In Cairncross' words, 'there are no definitions of the successive stages that admit of their identification by reference to verifiable criteria': and the balance of opinion appears to support a conclusion of this kind.

Criticism over the third criterion is even stronger. Although Rostow in fact claims that his stages 'are not merely descriptive' but 'have an inner logic and . . . analytic bone-structure' (Rostow, 1960, pp. 12-13), nearly all reviewers (Enke, 1964, p. 201; Drummond, 1961, p. 113; Cairncross, 1961, p. 451) agree that in fact his model 'simply fails to specify any mechanism which links the different stages' (Baran and Hobsbawm, 1961, p. 236) and 'is essentially an essay in classification' (Habakkuk, 1961, p. 601). Rostow himself places great stress on the concept of leading sectors, whose 'changing sequence' provides an essential raison d'être for his stage breakdown (Rostow,

1960, p. 14). But as Ohlin (1961, p. 649) points out, leading sectors are not even mentioned in detailed analysis of the first two stages; and discussion of their significance during take-off and post-take-off stages is insufficient to provide the model with the 'analytic bone structure' which Rostow claims for it.

The third level of criticism concerns the testing of the model against the real world. As Chorley (1964, p. 136) has pointed out, such testing is crucial to any final assessment of a particular model's value. Unfortunately, it is here that Rostow's model appears most open to criticism. Firstly, most economic historians (Ohlin, 1961, pp. 649–650; Hagen, 1959, p. 132; 1962, pp. 519–520; North, 1958, p. 75; Cairncross, 1961, pp. 454–456) deny that the histories of present-day advanced countries reveal any signs of a twenty-to-thirty year period in which investment rates suddenly rose sharply, denoting take-off. In Kuznets' words (1963, p. 35), 'the available evidence lends no support to Professor Rostow's suggestions' on this crucial point. Secondly, empirical observations have thrown serious doubt on the very separateness of Rostow's different stages. He himself admits that, for example, the age of high mass consumption can be coincident with the drive to maturity (see Fig. 8.3). However, most observers are agreed that historically, the preconditions and take-off stages are also often indistinguishable (Hagen, 1962, pp. 517–519; Kuznets, 1963, p. 37; Cairncross, 1961, p. 456; Habakkuk, 1961, p. 602). If this is so, it surely casts considerable doubt on the value of Rostow's model as an analytical and predictive tool.

Whatever the final verdict on the model, however, it must be admitted that it has stimulated an enormous amount of research into regularities in economic growth, ranging from empirical testing (Rostow, 1963) to the construction of mathematical models based explicitly on Rostow's ideas (Ranis and Fei, 1961). In addition, it has recently been used, apparently successfully, for setting economic growth in a particular country in a general context (Houghton, 1964). Such use suggests that it might well be valuable and illuminative as a teaching device, for comparing and classifying apparently different economies.

Rostow's is the only major example of a conceptual model which attempts to fit all countries. However, as Enke (1964, p. 204) has pointed out, one way of increasing its 'degree of fit' would have been to have framed it in terms of a smaller and more homogeneous group of state-units. A more limited approach of this kind is favoured by several economists (e.g. North, 1958, p. 75; Ruttan, 1959). The criterion of homogeneity envisaged, however, varies considerably. Enke himself suggests culture. Myint (1964, p. 36) on the other hand regards degree of overpopulation as the crucial factor, at least as far as the construction of models relating to underdeveloped countries is concerned. Hoselitz (1955A, pp. 417–418) favours a division between countries in which growth has occurred by an 'expansionist' process and those in which it has taken place by an 'intrinsic' one. Still other economists, notably Kuznets

4

(1960, p. 15), stress the importance of size, and the fact that partly because of economies of scale (Ewing, 1964, pp. 356–358; Chenery, 1960, pp. 645 and 651), partly because of different resource potential (Hicks, 1959, pp. 182–183), the nature and problems of economic development in a large country are often significantly different from those of a small one (Kuznets, 1951, pp. 29–31; 1953–54, pp. 14–16; and 1958; Deane, 1961, p. 18; Hoselitz, 1959, p. 145; Robinson, 1960). However, the only commentators who support their particular criterion by actual model construction are Fairbank, Eckstein and Yang (1960). They suggest a simple division of the world into more and less developed economies, and construct a five-phase historical model designed to fit most of the latter, including in particular India, China and other Far Eastern areas. Conceptually, the model clearly owes a vast debt to Rostow, being framed in terms of 'five phases characterized by (1) traditional equilibrium, (2) the rise of disequilibrating forces, (3) gestation, (4) breakthrough or as some prefer to call it, take-off, and (5) self-sustaining growth' (Fairbank, Eckstein and Yang, 1960, p. 1). However, in that its ideas apply specifically to one group of economies, the model's 'degree of fit' to individual cases such as China is clearly greater than that of Rostow's own general model. The development of more limited models of this kind referring to particular groups of countries might well prove a fruitful avenue for future research.

(2) *Mathematical models.* The construction of mathematical models of economic growth within national economies has become one of economics' fastest growth points in recent years. The reason for this lies not only in the growing realization of the value of such models for government planning purposes (United Nations, 1961; Hart, Mills and Whitaker, 1964), but also in the development of new statistical techniques and of computers capable of handling vast quantities of data and calculations (Orcutt, 1960; Cohen and Cyert, 1961; Mills, 1964; Stone, 1964A). Mathematical growth models contain at least three types of elements; components, which are the units for which data are collected (e.g. major industries, or individual households); variables, which describe some aspect of the components (e.g. labour force, or yearly expenditure); and relationships, which 'specify how the values of different variables in the model are related to each other' (Orcutt, 1960, p. 899; see also Tinbergen and Bos, 1962, p. 6). Once the elements of a particular model have been mathematically specified, usually in the form of a series of equations (Tinbergen, 1959; Enke, 1964, Chap. 9; Goldsmith, 1959, pp. 75–78), the implications of changes in any of them produced by growth can be calculated.

A simplified classification of mathematical growth models involves a division into 'aggregate' and 'inter-industry' models (United Nations, 1961). The former usually consist of a series of formulae, specifying relationships between variables such as production, consumption, investment, etc., within an entire economy. Good examples are the models developed by Klein (1961) for Japan, Mahalanobis (1963) for India, and Valavanis-Vail (1955) for the United States. Inter-industry models, on the other hand, are 'concerned with the quantitative analysis of the interdependence of producing and consuming units in a modern economy', and focus particular attention on the interrelationships among different producers 'as buyers of each others' outputs, as users of scarce resources, and as sellers to final consumers' (Chenery and Clark, 1959, p. 1). The simplest and commonest type is the 'input-output' model, developed by Leontief (1936; 1951A; 1951B and 1953). Input-output models group production activities in an economy into a number of sectors – e.g. in simplest form, agriculture, extractive industry, manufacturing industry, services, etc. (Chenery and Clark, 1959, pp. 13–15). These sectors are then listed as headings in a table, or 'input-output matrix', both down the left-hand side – where they are thought of as producing units – and along the top – where they are thought of as consumers. The spaces in each horizontal row of the matrix are filled in with the value of that sector's production (output) which is sold to each of the other sectors (as their inputs) during the period under study (Fisher, 1964, p. 20). An extra vertical column is usually included to contain values of output sold directly to consumers who are not themselves producers – i.e. to final consumers. From the values thus plotted, coefficients relating input from one sector to total output of another sector can be calculated (Leontief, 1965A, p. 33); and these, under certain assumptions (United Nations, 1961, p. 13), can be used to analyse the effects which growth in one sector – e.g. in total agricultural output – will have upon each of the others, and therefore on the whole economy. In other words, input-output models explicitly recognize that changes in production in one sector inevitably affect production in many other sectors as well – by means of altered demands for outputs from these other sectors as inputs to the original sector. Calculation of the full effects on other sectors is carried out by the 'iterative' method (Isard, 1960, p. 331), or 'round-by-round' computation of input requirements. Input-output models have proved very valuable for assessing the impact of changes of this kind, and for planning economic development generally in a number of countries (Barna, 1963, Part I; United Nations, 1961; Leontief, 1965B).

Another and similar type of model (Stone, 1956) is the 'social accounting' model. This also organizes its information in terms of a matrix, but interest is focused here on all the different kinds of monetary transactions (e.g. government payments, property income) which occur within an economy (Stone, 1956, p. 156). A social accounting model is at present being used by the Cambridge University Department of Applied Economics to analyse the sectoral implications of future economic growth in the United Kingdom (Department of Applied Economics, 1962; Stone, 1961; 1964B and 1965;

Hart, Mills and Whitaker, 1964). Finally, 'linear programming' models, representing an elaboration of input-output ones, are also now being used to analyse economic growth, and to determine 'the most economical way of achieving a given set of objectives' in economic development (United Nations, 1961, p. 14; see also Chenery, 1961 and 1963; Chenery and Clark, 1959, Chap. 4; Sandee, 1959).

Although non-spatial, most of the mathematical models discussed above can be used to throw light on the differences between countries and regions in nature and level of economic development, and it is because of this that they have been discussed here.

Sub-national-scale non-spatial models

Although most economists have studied non-spatial aspects of economic growth in terms of national units, a few (Gras, 1922; Robock, 1956; Isard and Smolensky, 1963; Fisher, 1955) have stressed the need for regional development studies and models, particularly in underdeveloped and very large countries. Hoover's conceptual model of regional growth (Hoover, 1937, pp. 284–285 and 1948, pp. 187–196), involving a shift in regional economic activity from self-sufficient subsistence agriculture, through agricultural- and mineral-based industries, to tertiary activity oriented to export markets, is very similar to that suggested by Fisher (1955, pp. 3–14). The key to both is regional industrialization, and the focus on the changing economic structure of the region is largely non-spatial. So too is Rostow's model which, as he points out (Rostow, 1960, p. 1), is capable of application at the regional level. Later conceptual regional models, however, have stressed far more the spatial relationships of developing regions with other areas in the same country, and will therefore be discussed in the context of Model Group B.

The development of non-spatial mathematical models of regional growth has proceeded very rapidly in the last few years (Meyer, 1963). Nearly all of these however represent only the application at the regional level of models, particularly inter-industry models, developed initially at national scales. Input-output growth models, for example, have in recent years been increasingly applied to regional as well as national economies (Barna, 1963, Part II; Isard, 1951 and 1960, Chap. 8; Isard and Cumberland, 1961; Maki and Yien-I Tu, 1962; Chatterji, 1964). Indeed the size of region involved has occasionally been as small as an urban area (Hirsch, 1959 and 1963). More general accounting models, too, have been applied at the regional scale (Jouandet-Bernadat, 1964; Hirsch, 1962; Hochwald, 1961): and simple forms of aggregate models have been used for planning regional economic growth, as in Pakistan (Haq, 1963). These regional models generally and necessarily take more account of external relationships than their counterparts on the national scale; but they are still largely non-spatial in character.

MODELS OF SPATIAL DISTRIBUTION
OF ECONOMIC DEVELOPMENT

As pointed out earlier, only in recent years have economists turned from their preoccupation with temporal variations in economic growth to consider in addition spatial variations in development (Hirschman, 1959, p. 144). Since then, however, the construction and testing of models of the spatial distribution of economic development has proceeded very rapidly. Indeed, even Rostow (1964, pp. 103 and 122–131) now stresses the major importance of such variation within countries, despite the fact that his earlier model almost entirely ignored it (Kindleberger, 1964, pp. 248–249). Since most work on spatial variation in development has been focused on the national scale (see Figure 8.1), this will be considered first.

National-scale spatial models

(1) *Regional income inequality models*. The existence of regional inequalities in income – in both absolute and per capita terms – within virtually all countries is by now well attested. Put another way, economic development is scarcely ever spread evenly over the whole area of a given state unit, but rather concentrated at certain points, producing a mosaic of regions at different levels of economic prosperity (Vinski, 1962; Hemming, 1963; Hirschman, 1958, p. 183; Paauw, 1961, p. 186; Milhau, 1956; Gannagé, 1962, pp. 62–65). It is true, as Klaassen, Kroft and Voskuil (1963, p. 77) have pointed out, that the range of spatial variation in level of development within a country will in large part depend upon the scale of regional sub-division adopted (see also Haggett, 1965B). But since spatial variation in development exists in most countries even on the simplest scale of sub-division – i.e. a 'north-south' division (see Eckaus, 1961; Paish, 1964) – this point does not affect the general statement above.

Once recognized, this kind of spatial variation has proved difficult to explain in terms of traditional models of interregional economic relationships, particularly trade models. Such models, based on the concept of static equilibrium, assume that given relatively free mobility of the factors of production – a situation which to some extent obtains within most countries (Harris, 1954, p. 369) – 'factor movements tend to bring about an equalization of income among regions' (Harris, 1957, p. 191). Any differences in income levels between regions must therefore be viewed as only temporary, due to some slight lag in adjustment. Unfortunately, however, as other authorities have pointed out (Borts and Stein, 1964, pp. 49–55; Borts, 1960; Myrdal, 1957B, p. 13; Williamson, 1965, p. 5; Bachmura, 1959, p. 1012; Sisler, 1959, p. 1100; Schuh and Leeds, 1963, p. 296), equalization models are of little use in illuminating the development of spatial variation in the real world, since such variation is not only remarkably persistent, but apparently increasing in many countries (Williamson, 1965, pp. 16–17; Economic Commission for Europe, 1955, pp. 143–144; O'Connor, 1963).

The inadequacy of equalization theories has prompted economists in recent years to put forward new conceptual models of the development of spatial variation in economic prosperity. Probably the most important of these is the model elaborated by Myrdal during the mid-1950's (Myrdal, 1956; 1957A and 1957B). Framed implicitly in terms of countries already populated but exhibiting a low level of economic development, Myrdal's

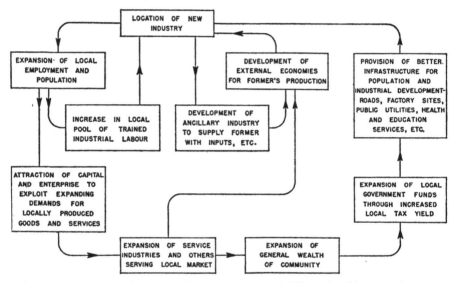

8.4 Myrdal's Process of Cumulative Causation – A Simple Illustration (the two sub-circles are included to illustrate the ramifications of the cumulative causation process).

model is based on the contention that in a free economy, particular changes do not, as equalization models contend, 'call forth countervailing changes but, instead, supporting changes, which move the system in the same direction as the first change but much further' (Myrdal, 1957B, p. 13). Applying this idea of 'cumulative causation' to the problem of economic development within countries, Myrdal concludes that 'the play of the forces in the market normally tends to increase, rather than to decrease, the inequalities between regions' (Myrdal, 1957B, p. 26): that is, that once particular regions have by virtue of some initial advantage moved ahead of others (see also Clark, 1966, p. 6), new increments of activity and growth will tend to be concentrated in the already-expanding regions *because of their derived advantages* rather than in the remaining areas of the country. The flow diagram of Figure 8.4 illustrates one possible example of this cumulative process (see also Pred, 1965, p. 165).

The concept of cumulative concentration is not however the only important feature of Myrdal's model. Closely involved with it in his explanation of differential regional growth is spatial interaction between the growing and stagnating regions. Once growth has begun in the former, Myrdal claims, spatial flows of labour, capital and commodities develop spontaneously to support it. Such flows operate, however, as 'backwash effects' upon the remaining regions of the country, since faced with the higher returns obtainable in the growth regions, these other regions tend to lose not only their more skilled and enterprising workers, but also much of their locally-generated capital. At the same time, goods and services originating in the expanding regions flood the markets of the remaining regions, putting out of business what little local secondary and tertiary industry may already have developed there. A further type of backwash effect operates through non-economic factors, such as the provision of poorer health and education services in the stagnating as compared with the expanding regions. In all these ways, backwash effects, particularly those working through spatial interaction, come into operation to frustrate growth in the former and sustain it in the latter (Myrdal, 1957B, pp. 27–31).

However, these backwash effects are not the only interregional relationships which the model postulates as developing within a growing economy. Also of significance are 'certain centrifugal "spread effects" of expansionary momentum from the centres of economic expansion to other regions' (Myrdal, 1957B, p. 31). By stimulating demand (e.g. for agricultural and mineral products) in other, particularly neighbouring, regions, expansion in the growing areas may initiate economic growth elsewhere. If the impact is strong enough to overcome local backwash effects, a process of cumulative causation may well begin, leading to the development of new centres of self-sustained economic growth. Such spread effects are however strongest in economies which have already achieved a fairly high level of economic development, since this 'is accompanied by improved transportation and communications, higher levels of education and a more dynamic communion of ideas and values – all of which tends to strengthen the forces for the centrifugal spread of economic expansion or to remove the obstacles for its operation' (Myrdal, 1957B, p. 34). At the same time, stronger spread effects will boost the economic growth of the country as a whole, by utilizing properly the resources of the formerly stagnant regions.

A basic assumption of Myrdal's model is government non-intervention in economic development. However, he does suggest that in advanced economies, stronger spread effects are aided by government policies aimed at fostering growth in backward regions, and that such action can be interpreted as only another aspect of cumulative causation in development. Certainly government concern with regional inequalities does appear to be greater in advanced countries such as the United Kingdom (National

Economic Development Council, 1963, pp. 14–29; Needleman and Scott, 1964; Humphrys, 1962 and 1963; Lonsdale, 1965; Manners, 1962; Fyot and Calvez, 1956), those of Western Europe (European Coal and Steel Community, 1961; Political and Economic Planning, 1962; European Free Trade Association, 1965; Barzanti, 1965; Romus, 1958; International Information Centre for Local Credit, 1964; Meyers, 1965; Ginsburg, 1957), and those of North America (Committee for Economic Development, 1958; Gilmore, 1960; Wood and Thoman, 1965; Wilson, 1964), than in the underdeveloped countries of the world, where government-inspired studies such as those of Yates (1961) and Furtado (1964) are few and far between.

Myrdal's interregional income inequality model, the first major example of its kind, was quickly followed by others. Hirschman, in his *Strategy of Economic Development* (1958, pp. 183–194) develops a model which although arrived at independently of Myrdal's is in many ways remarkably similar to it. The key role in differential growth is here accorded to spatial interaction between growing 'Northern' and lagging 'Southern' regions, in the form of 'trickling-down' and 'polarization' effects. As Hirschman agrees, these are the exact counterparts of Myrdal's spread and backwash effects (Hirschman, 1958, p. 187), involving just those movements of capital, labour and commodities which have been previously described. However, the model, far from assuming a cumulative causation mechanism, implies that if an imbalance between regions resulting from the dominance of polarization effects develops during earlier stages of growth, counter-balancing forces will in time come into operation to restore the situation to an equilibrium position. Such forces, chief of which is government economic policy, are not to be thought of as intensified trickling-down effects, but as a new element in the model, arising only at a late stage in development. Their inclusion, together with the exclusion of any cumulative mechanism (Hirschman, 1958, p. 187), represent the model's chief structural differences from that of Myrdal.

Yet another similar interregional growth model has been suggested by Hicks (1959, pp. 162–166). After stressing the significance of initial advantages in determining the exact location of growth regions, Hick's model follows Myrdal's in emphasizing the importance of derived, cumulative, advantages in their subsequent development. It also focuses attention on spatial interaction, and flows of goods, labour and capital between growing and lagging regions. However, these flows are interpreted only as representing the 'tendency for wealth to spill over' from growth centres (Hicks, 1959, p. 163), rather than as a hindrance to growth in lagging regions. The movement of labour, for example, is seen as aiding the latter, not hindering their expansion: while only capital transfers from growing regions to other parts of the country are examined – a direction of flow opposite to the main flows suggested in both Myrdal's and Hirschman's models.

Least elaborated of this group of interregional growth models is Ullman's

(1958 and 1960). In his formulation, the crucial variable is the 'selfgenerating momentum' (Ullman, 1958, p. 180) experienced by growth regions – a momentum which reflects the development in them of 'notable external economies of scale and the largest market in the country' (Ullman, 1958, p. 184). Spatial interaction, though implicit in his analysis, is not really examined, probably because he is far more concerned with sheer concentration of new development than with the creation of differences in per capita regional income.

The two most important of these interregional income inequality models – Myrdal's and Hirschman's – have attracted a great deal of attention since their formulation; and although much remains to be done, preliminary testing of these, both as models and against the real world, has proceeded far enough to justify certain interim conclusions. Firstly, the models, especially Myrdal's, do appear to possess a well-defined structure, in which important variables are clearly related to each other. Indeed, Myrdal's claim that his cumulative causation hypothesis could in principle be stated 'in the form of an inter-connected set of quantitative equations' (Myrdal, 1957B, p. 19) has partly at least been justified by an attempt at mathematical formulation (Singer, 1961). More important, Borts and Stein (1964, pp. 4–7) have shown by a simple mathematical model that 'Myrdal's views are logical; i.e. they contain no internal contradictions' (Borts and Stein, 1964, p. 4): while despite Paauw's criticism (1961, p. 186) of backwash/polarization effects as 'hardly a satisfactory explanation' of regional backwardness, most commentators (e.g. Friedmann, 1959, p. 174; Lasuen, 1962; Hughes, 1961) seem to agree that these do represent important variables in explaining differential development.

Testing and criticism of the models as they relate to the real world has concentrated on four main topics. The first of these is Myrdal's hypothesis of cumulative causation, a concept stressed in slightly different form by Hicks and Ullman. Many authorities (e.g. Economic Commission for Europe, 1955, p. 142; Chisholm, 1962, p. 159; Caesar, 1964, p. 238; Nicholson, 1965, pp. 164–169) appear to agree with O'Connor's comment (1963, p. 42) that in most countries 'economic development tends to be concentrated in the areas where most has already taken place'. From this proposition, it is only a very short step to Myrdal's, and the stressing of the 'cumulative advantage' of growth regions (Perloff and Wingo, 1961, p. 106), or of 'the cumulative process of growth in the concentration areas' (Economic Commission for Europe, 1955, p. 154). As a result, Myrdal's concept receives considerable support, both implicitly in studies which actually precede his (Royal Commission, 1939, pp. 29, 49 and 170; Sturmthal, 1955, p. 200), and explicitly in subsequent work (Spencer, 1960, p. 46; Wonnacott, 1964, p. 418; Cairncross, 1959, p. 109; Baer, 1964, p. 269; Lasuen, 1962, p. 188; Pred, 1965, pp. 160–166; Brandenburg, 1964, pp. 208–209). In most cases, commentators are concerned primarily with growth regions, and tend to stress as Myrdal does (1957B, p. 27) the role of external economies (Lasuen, 1962, p. 180), together with growth above certain 'thresholds' of economic activity (Pred, 1965, p. 165), and higher rates of technological innovation (Perloff and Wingo, 1961), in their cumulative expansion. The prospect of cumulative growth has even stimulated policy recommendations regarding concentration of government investment in underdeveloped countries (Mabogunje, 1965, pp. 436–438; Friedmann, 1963, p. 53; Lutz, 1960, p. 45).

The cumulative causation hypothesis appears therefore to have been accepted by most authorities as a relevant and useful concept in analysing spatial concentration of economic growth, at least during early stages of development. So, too, with perhaps one major exception, has the significance of backwash/polarization effects. The most important of these effects, specified in both Myrdal's and Hirschman's models, are interregional flows of capital, commodities and labour; and tentative conclusions about the existence and nature of these flows can be drawn from recent empirical studies. For example, it seems generally agreed that in many underdeveloped countries there does indeed occur a net capital transfer from lagging to growing regions. This pattern has been reported for Indonesia (Williamson, 1965, p. 7), with transfers from the outer to the central islands; Pakistan, with flows from East to West (Haq, 1963); Brazil, with private capital movements from the north to south (Robock, 1963, p. 108; Baer, 1964); Nigeria, with currency flows from 'the West, Mid-West, and North towards the East and Lagos' (Hay and Smith, 1966, p. 23); and Spain (Lasuen, 1962). Most of these observers (e.g. Haq, 1963, pp. 103–104) consider the pattern an important factor behind differential regional growth. The models' forecasts on commodity flows also tend to be supported by what little evidence exists. In Belgium (Verburg, 1964A, p. 144), Italy (Clough and Livi, 1956, p. 336; Eckaus, 1961, p. 314) and to some extent Britain (Smith, 1953, pp. 93–121), goods manufactured in the growth regions of the country during the eighteenth and nineteenth centuries helped to depress what industry already existed in the more backward areas. The same process may have taken place in Brazil in the nineteenth and twentieth centuries (Furtado, 1963, pp. 264–265).

In so far as the sheer existence of net population migration from lagging to growing regions is concerned, the Myrdal/Hirschman analysis of labour flows has been fully accepted. However, their views concerning the selectivity – in terms of age, ability or skills – of such migration, and its effect upon the lagging and growing regions, appear to be far more open to question. Some commentators (Hughes, 1961; Lasuen, 1962; Williamson, 1965, p. 6; Friedmann, 1959, p. 174; Parr, 1966, pp. 152–154) explicitly support them, as do certain other independent studies (Eckaus, 1961, p. 317; Hathaway, 1960; Randall, 1962, p. 78). However, Okun and Richardson (1961), in an important contribution, claim that the Myrdal/Hirschman argument on selectivity 'does not adequately deal with the complexities inherent in the relationships between migration and inequality of per capita income' (Okun and Richardson, 1961, p. 132). To demonstrate this, they develop a simple conceptual model which relates interregional migration within a closed economy to level and rate of economic growth in different regions. When applied to migration within the United States, this model does appear to show that the influence of migration – even age-selective migration – upon regional per capita income differences varies considerably according to the circumstances, particularly the direction and duration of movement. Okun and Richardson's conclusions therefore throw considerable doubt on the view of selective migration as a normal feature of backwash/polarization effects.

Empirical evidence concerning the third important aspect of these models – spread or trickling-down effects – is much scantier than that for backwash effects. Lasuen, for example, sees little sign of their operation in Spain, except in the immediate vicinity of Barcelona and Bilbao (Lasuen, 1962, p. 177): while Chenery's analysis (1962) of economic development in southern Italy shows that although the area's 'economic structure is influenced by being part of a more advanced economy' (Chenery, 1962, p. 517), this influence has operated far more through massive government income transfers to the South than through any natural spread of expansionary momentum from the North. This apparent absence of spread effects in many, particularly underdeveloped, countries is possibly partly due, as Hicks (1961, p. 77) has pointed out, to the considerable cultural, social and economic differences between regions. However, in more advanced countries where such differences are less, the geographical spread of economic development appears to have occurred partly by a mechanism ignored in the Myrdal/Hirschman model – that of industrial dispersal in search of labour, raw materials, or markets (McLaughlin and Robock, 1949; Sickle, 1951, p. 387; McGovern, 1965; Keeble, 1965; Cameron and Clark, 1966). In at least one case, this has been a major factor stimulating regional economic growth (Manners, 1964, p. 51). By contrast, governments in underdeveloped countries have generally found it very difficult to promote industrial dispersal (see Coutsoumaris, 1964, pp. 85–86). It can be suggested therefore, that although no general conclusion concerning the existence of spread effects is as yet possible, at least one further mechanism for the interregional spread of economic growth in advanced economies needs to be included in any analysis of them.

The final aspect of these models to have provoked considerable empirical testing is the question of divergence or convergence of regional per capita incomes. On its own, Myrdal's cumulative causation concept suggests continuing divergence in such incomes as a typical feature of developing countries. Hirschman's model, on the other hand, provides a theoretical background for regarding convergence as the norm. Unfortunately, however, empirical studies at first sight appear to arrive at conflicting conclusions.

Indeed, in the case of Brazil authorities are divided even as regards the same country, Baer (1964) and Furtado (Robock, 1963, pp. 107–110) claiming that the gap between regional incomes per capita has been widening, Robock (1963, p. 46) that it has been converging. Elsewhere the situation appears only slightly less confused. Disparity in regional incomes appears to have been increasing in Italy (Clough and Livi, 1956, p. 335), Mexico (Sturmthal, 1955, p. 201) and possibly Uganda (Elkan, 1959, p. 137), while convergence, of wage levels as well as regional incomes, seems to have characterized the recent history of the best-attested case, the United States (Smolensky, 1961, p. 68; Easterlin, 1958 and 1960; Fulmer, 1950, p. 273; Sickle, 1951, p. 389; Wonnacott, 1964).

This conflict is however more apparent than real; for both models in fact recognize the greater likelihood of divergence during earlier periods of growth, and of convergence during much later periods. In Myrdal's case, cumulative growth in lagging regions may well eventually be stimulated by a combination of spread effects and a stimulus external to his model, government intervention. In Hirschman's, divergence during an early stage is in fact anticipated, convergence only occurring later. This idea of a shift from divergence to convergence with increasing economic development – an idea referred to by other authorities (e.g. Balassa, 1961, p. 201) and paralleled by trends in personal income inequalities (Kuznets, 1955, p. 18) – has been taken up by Williamson (1965) in a major empirical investigation. His work shows that what little statistical information on regional income inequalities in different countries is available does support the view that poor but developing countries are characterized by increasing regional disparities, while more developed ones exhibit decreasing disparities. A statistical index of regional income disparities in any one country might thus trace out a path on a graph similar to one of those shown in Figure 8.5. An interesting point brought out in Williamson's study is the apparent secondary influence of geographical size upon degree of regional inequality. His analysis (Williamson, 1965, p. 15) lends some support to Kuznets' view (1960, p. 30) that 'the developed small states seem to have succeeded in spreading the fruits of economic growth more widely among their populations than the larger states at comparable levels of income per capita'.

Initial testing of the Myrdal/Hirschman models on the whole therefore suggests that they are congruent with reality – particularly the reality of present-day underdeveloped economies – in respects which make them valuable in its interpretation. However, this can only be an interim conclusion, and considerable further empirical work needs to be done before a full assessment can be made.

Although the interregional income inequality models just discussed have attracted most attention, a few other conceptual models may perhaps also be included under this general heading. For example, Keirstead (1948, pp. 265–

313) has put forward a model, based clearly on Canadian experience, of the process of regional depression and growth in an advanced country. Framed in terms of a two-region ('Eastland' and 'Westland') closed economy, this model suggests that the volume and nature of industry in these regions is normally in some kind of 'location equilibrium'; and that if technological changes favouring concentration of industry in the region (Westland) with the larger market upset this equilibrium, shifts in industrial location will occur 'until a new equilibrium in location, satisfying the new conditions, is achieved' (Keirstead, 1948, p. 269). The creation of a labour cost differential favouring the depressed region plays a major part in this adjustment process.

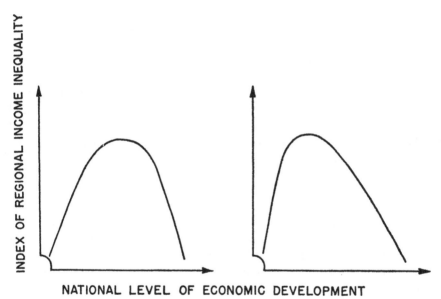

8.5 Possible Relationships between Indexes of Regional Income Inequality and National Development (*After Williamson, 1965, pp. 9–10*).

Two further conceptual models have been suggested by Friedmann. The earlier (Friedmann, 1956) represents an attempt to relate certain insights of location theory, such as the concept of functional hierarchies of cities and city regions, and of agglomeration economies, to the question of spatial changes associated with economic growth. As economic development proceeds, Friedmann claims, the degree of areal specialization, functional differentiation and spatial interaction between different regions increases. As a result, the economy's spatial structure develops from one characterized by 'small, isolated, and functionally undifferentiated communities', into a hierarchy of interdependent regions, this giving way at the highest stages of development to 'more or less "autonomous" linear cities and conurbations of very large proportions' (Friedmann, 1956, p. 226). His later model (1963) relates specifically to Latin American countries, and is couched in terms of four phases of development – initial coastal settlement; the phase of semi-autonomous, externally-orientated regions; industrialization and development of a centre-periphery structure; and final integration of the national space economy through the spread of metropolitan regions. Although bringing together many useful ideas, neither of these two models throws much new light on spatial inequalities in development, or has been followed up by empirical work.

(2) *Export base models.* Regional income inequality models are not, however, the only intellectual constructs to have been developed for studying variations in regional economic growth. A different approach has been followed by certain economists concerned with regional economic history and growth in the United States. Impressed by historical evidence, these have stressed the key role of a region's 'export base' – that is, 'collectively the exportable commodities (or services) of a region' (North, 1955, p. 248) – in determining the rate of its economic growth. Indeed, Perloff and Wingo (1961, p. 200) claim that in the United States 'regional growth typically has been promoted by the ability of a region to produce goods or services demanded by the national economy and to export them at a competitive advantage with respect to other regions'. This view has led such writers (e.g. North, 1955; Thomas, 1963) to abandon earlier regional growth models (see page 256) and to develop ones stressing the role played by the export base. North (1955), for example, has put forward a five-stage regional export base model which he suggests can be applied not only to the United States, but also to other capitalist countries in which population pressure has been slight. After a very brief subsistence stage, he envisages the rapid development of exporting of staple commodities to more advanced regions as the basis of the regional economy. With the growth of external economies, inflow of capital, and provision of an export-orientated infrastructure, a further stage of export intensification and regional development ensues, leading in time to the development of 'residentiary' industry, serving local markets. Finally, the expansion of residentiary industry, together with 'footloose' industries located more or less by chance in the area, may reach a point at which they too enter export markets, thereby diversifying the region's export base. This model is closely similar to ideas put forward by Perloff and Wingo (1961, pp. 200–201), and shares a common frame of reference with Baldwin's analysis (1956) of the role of the production function of the main export commodity in influencing the development of other sectors of economic activity in a newly-settled region.

North's model focuses attention on the mechanism of economic growth of one particular region. Duesenberry's export base model (Duesenberry, 1950, pp. 96–102), however, is more concerned with the impact of economic growth in a newly-settled, exporting region, upon that of an older, already developed region. Clearly based upon the historical experience of the East and West North Central regions of the United States, this model suggests that in addition to 'western' economic growth consequent upon the development of exports of farm products, 'eastern' economic growth will also be stimulated by increased demand from the former region for more sophisticated products (e.g. manufactured goods) which are not as yet produced in the west. Given a sufficient rate of western agricultural expansion, growth induced by this demand will be more than enough to offset local agricultural decline due to increased western competition. Finally, rise in income in both regions will, as in North's model, generate expansion in local industries and services to serve expanding local markets.

Duesenberry's two-region export base model has attracted little comment, although later studies of the United States (Conrad and Meyer, 1965, pp. 225–226), Canada (Meier, 1953, p. 5) and Mexico (Glade and Anderson, 1963, p. 77) lend some support to its analysis of growth relationships between older, developed regions, and newly-settled, 'frontier' areas. The more general export base model exemplified by North's analysis, however, has stimulated considerable controversy concerning both its conceptual framework and empirical validity. Tiebout's criticisms (1956) are largely in terms of the former. He claims that variables other than exports – such as 'business investment, government expenditures, and the volume of residential construction' (Tiebout, 1956, p. 161) – affect regional income, and that the development of residentiary industry is of great importance to a regional economy in lowering costs in export industries. Tiebout (1956, p. 161) also stresses that 'the quantitative importance of exports as an explanatory factor in regional income determination depends, in part, on the size of the region under study' – a point on which Harris (1957, p. 169) is in agreement. North's immediate reply (1956) concedes the point about residentiary industry, while his subsequent article (North, 1959) admits that the diversion of income received from exports into other activities within the region plays a more important role in sustained regional growth than he formerly allotted to it. This slightly amended model provides the conceptual framework for his major analysis (North, 1961) of nineteenth-century economic growth in the United States.

Although Wilhelm's earlier discussion (1950) of regional economic growth in Soviet Central Asia provides considerable support for the export base model, explicit testing has been confined to examination of United States' experience. An important contribution here is that of Borts (1960), whose analysis of factors influencing changing regional wage levels in the United States 'indicates strong support for a model of regional growth based on the demand for a region's exports' (Borts, 1960, pp. 342–343). Detailed study of one particular region, the Pacific Northwest, confirms this view for that area (Tattersall, 1962), while other studies (Perloff, 1960; Perloff and Wingo, 1961; North, 1961) do appear to demonstrate the relevance of the model to the economic history of other parts of the United States. The most recent of these, and one of particular interest, is Borts and Stein's sophisticated analysis (1964). These authors use mathematical growth models to explore changing economic relationships during regional economic development, and accord export activities, and changing demand for regional exports, a major role in explaining regional growth differentials in the United States (Borts and Stein, 1964, pp. 121 and 132). However, their analysis also places considerable weight upon another variable in regional economic growth – the role of regional differentials in wage levels, and related changes in labour supply. Both wage levels and exports are held to be of equal importance since either or both of them may stimulate 'an expansion of investment in the region' (Borts and Stein, 1964, p. 121) and regional economic growth. This conclusion is reinforced by careful statistical comparison with United States experience, and suggests that although 'supported by considerable evidence' (Friedmann and Alonso, 1964, p. 210), the simple export base model may well now need refinement if it is to remain a useful tool in the comprehension of regional economic growth.

(3) *Mathematical models.* The development of mathematical models focusing explicitly on variations in the spatial distribution of economic development within a given state-unit has lagged far behind that of the non-spatial models discussed earlier. Perhaps the simplest work along these lines is represented by the use of regression analysis to determine the degree of statistical correlation between level of economic development and other variables presumed to influence this, in different regions of a given country. Fulmer's study (1950) is an early example of this, while Thompson and Mattila (1959) use more sophisticated multiple correlation techniques in their study of the factors associated with rapid employment growth in different states of the United States. In one sense, these studies do represent attempts at model-building, since the end-product is the specification of a generalized relationship between certain, possible causative, variables, and level of economic development. Indeed, in Klaassen, Kroft and Voskuil's study (1963), the use of regression analysis is closely integrated with the development of an actual mathematical model, constructed to throw light on the factors behind differences in regional per capita income in Holland.

This particular model developed by Klaassen, Kroft and Voskuil is an aggregate model, relating such variables as size and rate of natural increase of the working population, economic structure, etc., to regional economic development. As such, it illustrates the way in which aggregate models, although originally non-spatial (see page 254), are now also being developed explicitly for the purpose of illuminating regional differences in economic growth within a given state-unit. Other examples of this trend are the aggregate models developed by Borts and Stein (1964) to explore variations in regional development in the United States, and that put forward by Klaassen in his recent study (1965, pp. 43–49) on area economic development. This latter model takes the form of seven equations, relating such variables as demand for labour in 'basic' and 'non-basic' industries (see Alexander, 1954), net migration, natural growth of the working population, and the wage or income differential between the area concerned and the outside world. When certain of its predictions are tested against actual per capita income data for the regions of Belgium, a reasonably close fit is obtained, justifying Klaassen's claim that in spite of their simplicity (Klaassen, 1965, p. 38), such aggregate models 'can help considerably to understand why and how differences in income level arise between areas and regions' (Klaassen, 1965, p. 43).

Though they do not fit neatly within the model classification adopted here, Markov chain models have also been used to illuminate the question of differential regional economic growth. In his pioneering study, Smith (1961) examines the implications of increasing the trade of a depressed region under equilibrium trading conditions, in terms of a stochastic or Markov matrix of transition probabilities. Given certain limiting theorems of Markov chain theory, Smith's analysis clearly indicates that a permanent increase in the depressed area's income relative to other regions can only be achieved under equilibrium trading conditions 'by continuous injections' of income. The application of such Markov chain models to the problems of regional economic development seems bound to increase.

The most developed form of mathematical model relating to differential regional economic growth, however, is the inter-industry model. Such models, particularly input-output ones, are easily adaptable to interregional analysis, as Isard (1951 and 1960, Chap. 8), Leontief (1953, Chaps. 4 and 5) and Chenery (1956) have shown. For 'n' regions, the full list of 'x' sectors into which each regional economy is divided for input-output purposes is repeated 'n' times, both down the left-hand edge and along the top of the matrix. This enables inter-sectoral flows to be recorded not only within each region, but between different sectors in different regions (Isard, 1960, p. 318). Alternatively, separate input-output matrixes may be drawn up for each region, flows from different sectors in other regions being recorded in the form of supply coefficients in columns in the right half of the matrix (Chenery, 1956, pp. 345–347). As with non-spatial input-output matrixes, coefficients relating flow of inputs from any given sector of any region to another sector of any region are derived from recorded transactions.

The usefulness of input-output models for analysing the locational implications of economic development is considerable. For example, they are particularly valuable for determining the impact of increased investment or other expenditure in one region upon economic activity in others. Chenery (1956, p. 353) and others (Chenery, Clark and Cao-Pinna, 1953) have applied interregional input-output models to the Italian economy for just such a purpose. Again, they may be used to discover the optimum locational pattern of new investment for the achievement of 'balanced interregional growth' (Tinbergen, 1960, pp. 8 and 12): or to discover the minimum investment outlay required in each region to achieve given regional income targets (Tinbergen, 1964, pp. 2–7). In fact, as these examples suggest, interregional input-output models have been developed primarily as analytical and predictive tools for government planning of economic development – an association clearly illustrated in two important recent publications (Barna, 1963, Part II; Isard and Cumberland, 1961, pp. 287–338) referring to them. The use of these models in government planning has, it is true, thrown up certain problems. One is the enormous difficulty of obtaining the detailed regional data needed for input-output analysis from any but the most advanced countries. Another, perhaps more important, is the failure of models framed in terms of constant interareal input coefficients (Isard, 1960, p. 333) to account for the effects of scale and external economies available in particular regions (Isard and Smolensky, 1963, p. 109; Isard, 1960, pp. 338–343). However, even with these problems, inter-industry models have proved very useful in assessing the locational implications of government plans for economic growth; and with the refinement of such models, as for example in the direction of interregional linear programming techniques (Isard, 1958 and 1960, Chap. 10; Chenery, 1963; Stevens, 1958; Berman, 1959), their application for policy purposes to the analysis of interregional differentials in economic growth seems bound to increase.

Supra-national-scale models

The remarkable postwar growth of interest in the mechanism and characteristics of economic development referred to earlier has partly been stimulated by the apparent steady increase in recent years of income disparities between state-units. Many authorities (Andic and Peacock, 1961; Kuznets, 1953–54 and 1956; Briggs, 1965; Myrdal, 1957B, p. 6) support Deane's contention (1961, p. 16) that as far as income per capita is concerned, 'international inequality is increasing and that it is appreciably greater than the intranational inequalities'. However, although at first sight well documented statistically, this conclusion needs to be qualified in at least two ways. Firstly, as a number of writers have pointed out (Usher, 1965; Rao, 1964, pp. 66–104; Kindleberger, 1958, p. 3; Hagen, 1960, pp. 63–64), apparent income differences between different countries may well be 'equalizing' – i.e., the real value of a given amount of income in an apparently poorer country may be much greater (because of lower prices for goods and services) than that of the same amount in an apparently richer country. This tendency must reduce the apparent income gap in many cases, particularly as between industrial and underdeveloped economies (Rao, 1964, p. 103). Secondly, however, a widening gap in per capita incomes between countries may occur statistically even when the poorer country's economy is expanding faster than the richer one's. This is because, as Myint (1964, p. 18) points out, 'arithmetically, the widening gap depends not only on the differences in rates of growth but also on the initial width of the gap'. Again, this suggests that a widening gap should not cause too great a concern, since considerable economic growth may none the less be occurring in the poorer country concerned.

These two qualifications are important. However, they are clearly insufficient to justify ignoring the observed growth in international income disparities over recent years. This growth runs counter to the theoretical predictions of static equilibrium analysis, which suggests that trade stimulates economic development in all participating countries, and in the long run helps to bring about equalization of per capita incomes (Haberler, 1959; Cairncross, 1962, Chap. 13; Neumark, 1964; Caves, 1960, p. 259). This discrepancy between traditional assumptions and observed reality has led certain economists in underdeveloped countries to discard static equilibrium models, despite their internal consistency (Das Gupta, 1965, p. 123), and to put forward other ideas about the nature of spatial interaction between poor and rich countries. Probably the most influential of these economists is Prebisch (1950 and 1959), who has developed what might be called a 'centre/periphery' model of international economic development. Although on the world scale the 'contour lines of international economic inequality' (Briggs, 1965, p. 15) clearly identify the centre as the zone of highly developed economies stretching from European Russia to the United States and Canada, the terms 'centre' and 'periphery' are, as Friedmann and Alonso point out (1964, p. 211) 'more than a description of geographic position'. They further connote 'a set of structural relations that hold the periphery in nearly permanent subordination to the urban-industrial heartland' (Friedmann and Alonso 1964, p. 211). Chief amongst these relations are the enforced exploitation of natural resources in peripheral countries, at least during earlier colonial periods; the flooding of peripheral markets with manufactured goods produced in the countries of the industrial centre; and the unfavourable secular trend in international terms of trade as far as the primary, chiefly agricultural, goods which are produced at the periphery are concerned (Prebisch, 1950; Friedmann, 1963, pp. 44–45; Meier and Baldwin, 1957, p. 147). That such production characterizes the periphery, and is a response to the demands of the centre, is verified by Melamid (1955), in terms of Von Thünen's model of agricultural location.

This centre/periphery model, as Friedmann (1963) has pointed out, is applicable not only on the world but on the continental scale. For example, United Nations experts (Economic Commission for Europe, 1955) have drawn attention to just such a spatial pattern of per capita income inequality within Europe itself (see Figure 8.6). They find not only that 'the countries situated near the economic centre of Europe are, in general, richer and more developed than those at the periphery', but also that within European coun-

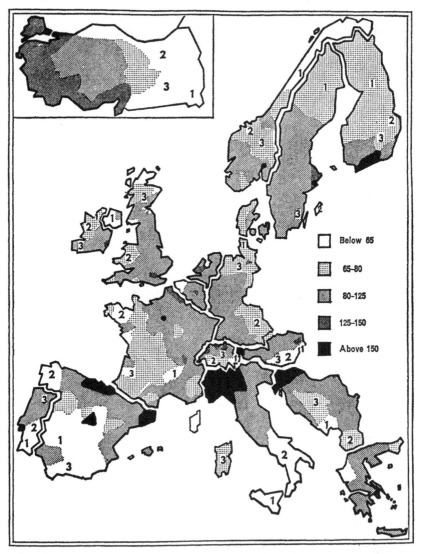

8.6 Regional Income Disparities within West European Countries (shading indicates estimated per capita income in each region expressed as a percentage of national income in each country: *Source: Map 2, Economic Commission for Europe, 1955*).

tries 'the levels of economic development tend to be lowest in the regions furthest removed from the relatively small area which developed as the main European centre of industrial activity, embracing England and the valley and outlet of the Rhine' (Economic Commission for Europe, 1955, p. 138). Of course, the restrictions on regional economic growth imposed by locations which are peripheral within particular European countries have by no means yet fully disappeared (see Verburg, 1964B). But on the whole, the supranational pattern is supported by national-scale studies, too, such as that by Vinski (1962, p. 138) on Yugoslavia. Again, this pattern seems to reflect a set of economic relationships which restrict economic development at the periphery, by, for example, 'preventing the spreading of industry to the poor regions' (Economic Commission for Europe, 1955, p. 154), in the kind of way suggested by Prebisch.

Prebisch's model, and the ideas put forward in the Economic Commission for Europe's survey, can however be viewed as only a special case of Myrdal's or even Hirschman's income inequality model; that is, the case in which backwash/polarization effects are dominant. The authors of these two latter models in fact explicitly extend them to the supra-national scale, where Myrdal's thinking in particular closely parallels that of Prebisch – 'internationally, however, the backwash effects of trade and capital movements would dominate the outcome much more, as the countervailing spread effects of expansionary momentum are so very much weaker' (Myrdal, 1957B, p. 54). However, the inclusion in Myrdal's analysis of the cumulative causation mechanism (e.g. Myrdal, 1957B, p. 19: see also Das Gupta, 1965, pp. 119–120), together with an acknowledgement of the operation of some spread effects, though on a very limited scale, does provide his model with a more flexible and meaningful structure than is possessed by the earlier centre/periphery analysis.

This latter point on spread effects is important in that it allows consideration of the significance for economic development in a poor country of spatial proximity to a richer one – development which empirical evidence suggests is sometimes aided by certain spread effects from the latter. An inference of this kind could be drawn merely from the visual evidence of Figure 8.6. Firmer support, however, is available from empirical studies of particular countries. Glade (Glade and Anderson, 1963, p. 15), for example, has pointed out that as far as Mexico is concerned, 'without doubt, too, the geographical proximity of the United States, combined with tourism and the bracero movement, has facilitated the cross-cultural transmission of new values, tastes and attitudes' necessary for economic development; while at the same time, 'geographical proximity to the large and growing United States markets' has greatly aided the expansion of Mexican exports of primary products (Glade and Anderson, 1963, p. 16). Glade might also have pointed out that the massive annual inflow of foreign currency from American tourists, and the considerable remittances from temporary migrant workers – braceros – in the United States, are themselves largely a function of the geographical proximity of the two countries, and represent a spilling-over of economic prosperity from the United States to Mexico of great value for the latter's economic development. A second area benefiting in this way is Bechuanaland. In his recent study, Munger (1965, pp. 38–39) stresses the considerable advantages enjoyed by the Protectorate because of its 'physical proximity to South Africa, and especially the booming Witwatersrand'. For example, 'it is hard to realize the savings in money, time and general efficiency of having nearby specialists and research stations'; while proximity permits temporary migration of many Bechuanas to South Africa, and the remitting of wages back to the Protectorate (Munger, 1965, pp. 86–87). This latter phenomenon is not, it is true, specified as a spread effect in Myrdal's original model, but may surely here be considered as one, particularly in view of Okun and Richardson's conclusions discussed earlier.

The above evidence supports the contention that some spread effects do operate today on the supra-national scale, even if only between neighbouring countries. To that extent, Myrdal's model appears more useful in analysis of spatial interaction in economic development between states than does the centre/periphery one. This is probably even more true if international differences in economic growth during earlier centuries – particularly the nineteenth – are the subject of study. Reversing the centre/periphery model, in fact, Nurkse (1961, p. 14) has claimed that during the nineteenth century 'a vigorous process of economic growth came to be transmitted from the center to the outlying areas of the world' by the mechanism of trade in primary products – that is, exactly one kind of mechanism specified by Myrdal (1957B, p. 31) as a possible spread effect. Again, Kuznets seems to be suggesting in one or two publications (e.g. 1953–54 and 1959) that international differences in economic development in past centuries must be viewed within the framework of the 'gradual . . . and uneven spread to other countries' (Kuznets, 1953–54, p. 18) of the 'industrial system', and the economic prosperity which it brought. This rather special kind of spread effect did not, of course, necessarily influence all countries located close to the birthplace of the system, Great Britain. More important for successful transmission were 'similar material conditions' in the poorer country, or close association 'by social and cultural antecedents' (Kuznets, 1953–54, p. 22) – conditions met, for example, in New England, which early adopted factory production of textiles on the basis of skills and techniques imported directly from Great Britain, rather than in Portugal, where little penetration of the industrial system occurred. Whatever the direction of transmission, however, the stimulus which this spread of industrial technology and attitudes gave to sustained economic growth in some countries strongly supports its inclusion in any discussion of international differences in economic growth during the nineteenth century; and this in turn lends added weight to the conclusion regarding the usefulness of Myrdal's model reached above.

The use of regression analysis and other statistical techniques to determine the degree of statistical association on the supra-national scale between variations in level of economic development (usually measured in per capita income terms), on the one hand, and variations in such factors as total output of manufacturing industry (Chenery, 1960), size of state-unit (Kuznets, 1953–54), city size distributions (Berry, 1961A), energy consumption (Schurr, 1965), and latitude, climate, political status and economic structure (Berry, 1960 and 1961B), on the other, has expanded rapidly in recent years. The role that geographers, especially Berry, have played in this is extremely important, and the work done of great interest. However, actual mathematical models of the growth of differences in economic development between state-units have not yet been developed to any great extent. Leontief's recent use (1965B) of input-output models to compare the economic structure of countries at different levels of development perhaps comes close to it, particularly since he shows how to develop a hypothetical input-output model of one underdeveloped economy (Israel) 'as it would appear if it enjoyed self-sufficiency' (Leontief, 1965B, p. 140), and how to use such a model in assessing the changes needed to achieve higher levels of economic development in different countries. Obviously, too, interregional input-output models can also be used on this scale, to investigate the impact of changes in final demand in one country upon the economy of another. Generally, however, mathematical models have been developed for analysis of economic development at national and sub-national scales, rather than to account for spatial interaction and areal differentiation in development between state-units.

Sub-national-scale models

(1) *The regional multiplier concept.* In analysing the development of spatial variation in economic prosperity within regions of countries, the regional multiplier concept provides an important starting point. This concept has already been briefly introduced in discussion of the export base model of regional development (see page 266). Indeed, that model is in one sense incomplete without it, since the operation of a multiplier process following the inflow of export earnings to the region is essential to overall regional growth. However, in that the focus of attention in multiplier analysis is internal change within the region, discussion of the concept, though intimately related to the export base model, appears to fit more logically within this sub-national scale group.

Broadly speaking, the regional multiplier concept concerns the way in which a rise in income, production or employment in one group of economic activities in a region stimulates the expansion of other groups, through an increased demand from the former group and its workers for the goods and services produced by the latter. This rise is typically induced by changes external to the region. The inter-industry stimulus may take the form of an expansion of demand for actual production inputs needed by the original group; or it may operate indirectly through growth in demand for consumer-oriented goods and services from the better-paid and/or increased number of workers employed by that group. The activities which benefit in the latter case are of course the 'residentiary' industries, primarily engaged in serving local markets, referred to earlier (see page 266) – and these will also gain from the expansion induced in industries producing inputs for the original group, since their workers will also presumably generate a larger demand for local goods and services.

One of the first economists to recognize the importance of the multiplier process in local economic growth was Barfod, in his study (1938) of the economic impact upon the city of Aarhus, Denmark, of a large local oil factory, Aarhus Oliefabrik, A/S. Starting with local payments, in the form of wages, salaries and payments to local input suppliers, Barfod demonstrated that what he termed the 'primary elementary income' derived by the area from the factory was somewhat less than total local payments, owing to income leakages from the local economy. This kind of effect is illustrated by Figure 8.7 for the case of payments to local suppliers. However, he then showed that such 'primary investment or expenditure calls forth a series of secondary effects as links in a geometrical progression' (Barfod, 1938, p. 39), since part of this primary income would again be spent locally, thus providing further local wages and income, part of which would again be spent locally, and so on, until these secondary effects were exhausted, As a result, the total local income derived from total payments made locally by the Aarhus factory (based upon figures for 1937) was in the final result 27 per cent greater than the original sums paid (Barfod, 1938, p. 55).

This pioneering study was primarily concerned with elucidating the existing structure of income flows between one economic activity and the rest of the economy within which it was located. Daly, however, in his slightly later analysis (1940) of interwar employment growth in southern and midland England, not only adopted the term 'geographical multiplier', but stressed the now-accepted view that this should be 'a dynamic concept', concerned not with 'a static relationship between the unimpeded and localized industries' but with 'the effect of a change in the numbers employed in one type of industry upon the magnitude of the numbers employed in the other' (Daly, 1940, p. 250). Daly's 'unimpeded' industries were those which developed in an area for reasons unconnected with serving local markets – i.e., logically, export-oriented industries; while his 'localized' industries were those serving local needs, such as building and service industries – i.e. virtually the same as North's residentiary industries. Assessment of the multiplier effect therefore involved defining these two kinds of activity in some way and measuring over some period of time the ratio of employment growth in the former to that in the latter. For Britain over the 1921–31 period, the ratio was 1 : 1·042.

Probably because of its predictive possibilities, Daly's approach rather than Barfod's was the one adopted by subsequent multiplier studies. For example, Vining (1946, p. 203) also divided regional economic activity into two groups – 'carrier' industries and 'passive' industries – representing export and resi-

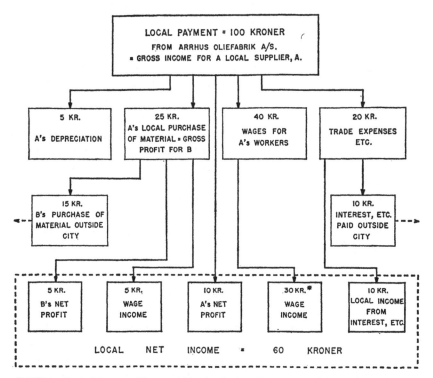

8.7 Hypothetical Income Flows within the Aarhus Economy (*After Barfod, 1938, p. 31*). Barfod reduces the contribution of wage income to local net income (*) by a factor of ·75, to allow for loss of the former, presumably outside the town.

dentiary industries respectively; and he too derived a multiplier ratio – of about 1 : 2 – for employment growth in these industries in one particular area, the Pine Bluff trading region of Arkansas. Hildebrand and Mace (1950), too, in their multiplier analysis of industrial growth in Los Angeles, built explicitly on Daly's work. Concerned with predicting employment growth in this rapidly expanding area, they classified industries as 'non-localised' or 'localised' according to their 'location quotient', an index of the degree of spatial concentration in a given area of particular industries (see Florence, 1948; Isard, 1960, pp. 123–126). From employment data, they then calculated a multiplier value of 1:1·248 for the period from just before to just after the Second World War. Their stress on the short-term applicability of this value was fully justified by their own calculations, which revealed a massive difference between the early (1:0·929) and later (1:2·785) parts of this period, if separate multiplier values are obtained for them.

These early multiplier studies, and associated 'economic base' analyses (see Andrews, 1953–56; Lane, 1966), were valuable in focusing attention on the importance of internal relationships in overall regional growth. However, as Isard (1960, p. 204) has pointed out, 'a regional multiplier derived from a basic-service ratio . . . has a strictly limited degree of usefulness and validity'. A full analysis of its shortcomings has been made elsewhere (Isard, 1960, pp. 194–205), but it is worth stressing two important ones. The first is the practical difficulty of separating export (basic) from residentiary (service) industries. Whatever method is used for such classification, no fully satisfactory result seems possible, since in any region some if not many industries ship large quantities of their products to both extra- and intra-regional consumers. The use of the location quotient for classification is open to serious criticisms (Isard, 1960, pp. 125–126). Secondly, multiplier analysis of this kind, particularly that based on a division by means of the location quotient, tends to ignore the direct kind of stimulus to other industries which increased demand for inputs from a growing export industry occasions (see above). This is because many of the former industries will, if already geared to the latter, probably possess a high location quotient, and therefore will have been classified in the export rather than residentiary sector. The full effect on regional economic growth of expansion in one particular export activity is therefore very difficult, if not impossible, to assess by this means (Isard, 1960, pp. 201–203).

For these and other reasons, later writers have on the whole adopted a different approach to multiplier analysis. This involves the use of input-output models. Such models can easily be adapted to focus on the changes which growth in one industry will stimulate in all the other industries of a given region. Probably the earliest study to attempt this was Isard and Kuenne's analysis (1953) of the direct and indirect repercussions upon local production and employment of the location of a new integrated iron and steel plant in the Greater New York – Philadelphia industrial region. Beginning with an estimate of production and employment in the new steel plant, the authors then calculated the probable growth of directly related steel-fabricating industries – this on the basis of historical trends and empirically derived indexes of association. The next step, involving the use of an existing matrix of input coefficients, was the determination (in value terms) of the first round of inputs from all other sectors needed for full-scale production of both steel and the products of the expanded steel-fabricating industries. The impact upon the household sector was included here. After reduction to allow for inputs produced outside the region, the second round of inputs needed to produce this first round of output expansions was calculated – and so on, until the increase represented by the 'n'th round was negligible. The total value of production, and corresponding employment, likely to be stimulated in the Greater New York—Philadelphia industrial region by the new steel plant could then be determined by summing the totals of each round. In this case, an extraordinarily high multiplier value – 1:19·66 – was obtained, due largely to an anticipated clustering of new or expanded steel-fabricating plants in the area, and their additional multiplier effects. Subsequent discussion of this result and method (Moses, 1955; Kuenne, 1955) added little of value, although Isard's later comments (1960, pp. 357–358) do point out certain, probably compensating, shortcomings.

Isard and Kuenne's study showed the value of input-output models for multiplier analysis, particularly when the impact of growth in one particular industry needs to be examined. Later studies have also utilized such models for this purpose. In their work on Utah, Moore and Petersen (1955) used an input-output model to assess the impact of variations in external demand for the products of Utah's main export industries upon the state's economy. Not only did this involve estimation of multiplier effects, but multiplier values were used as one index of the actual importance to the regional economy of different export industries. Moore's further work (1955) on Utah and California also assessed multiplier effects by an input-output model, and is interesting because it acknowledges that such effects 'may arise either internally or externally' (Moore, 1955, p. 136). That is, that both growth in an export industry, and, for example, internal population expansion, may produce a multiplier effect on regional income and employment. That the latter kind of change is also important in this respect is evident from empirical study of an area such as California. As a final example, Hirsch's work (1963) on St. Louis has demonstrated the value of input-output models for determining multiplier effects even within very small regions – in this case, urban regions. Such studies, potentially at least, are of great relevance for forward planning by local government authorities, and the use of input-output models in this way seems bound to increase.

Although the main approaches to regional multiplier analysis have now been outlined, a final comment concerning a topic often associated with it seems necessary. This topic is the so-called 'threshold' concept, which takes as its starting point the view that 'the size of a region's markets has an important bearing on the outcome of its economic development' (Pfister, 1963, p. 151). Clearly, this is true in that the larger the regional market, the greater the volume of local residentiary industry that can be supported. However, it is even more true in that for many manufacturing industries, economies of scale dictate a minimum size of plant (Pratten and Dean, 1965) and therefore a minimum size of regional market capable of supporting the particular

industry concerned (Guthrie, 1955). This minimum market size is termed a
'threshold' (Pred, 1965, p. 168), and protagonists of the concept hold that
attainment of a certain market 'threshold' is essential if regional growth is to
become self-generating (Tattersall, 1962, p. 229). Clearly, for individual
industries, the idea of a particular threshold is often justified. For example,
the building of petroleum refineries at Seattle almost certainly reflects the
growth of the Pacific Northwest regional market to a size which can just
support such activity (Pfister, 1963, p. 151). However, as Tattersall (1962,
p. 229) points out, the threshold concept is far less valid if applied to a region
as a whole, in the sense that attainment of one particular, crucial, level of
regional demand suddenly initiates a great leap forward in regionally-
oriented industrial activity. This view, implicit for example in Pfister's
comments (1963, p. 151) on the Pacific Northwest, and Perloff and Wingo's
argument (1961, p. 216) on regional growth in the United States generally, is
rightly criticized by Tattersall (1962, p. 229) on the grounds that 'accretion
of this type of industry is much more gradual than the "threshold" concept
suggests', and that there 'seems no reason why a "threshold" should be
achieved in all, or even many, industries at about the same time'. Only when
applied to individual industries can the threshold concept really be justified.
However, the existence of a threshold for different industries ought certainly
to be recognized when regional input-output multiplier analysis is being
carried out, since, for example, the attainment of such a threshold as a result
of multiplier effects may well alter the assumptions on which the input-
output analysis is based – e.g. as regards proportion of regional demand
satisfied by internal as compared to external producers. Indeed, input-
output multiplier analysis might even be used to assess the scale of expansion
in, say, an export industry, which is needed to stimulate growth to the
threshold level of regional demand for the products of another, potentially
residentiary, activity.

At the beginning of this section, it was suggested that regional multiplier
analysis provides a valuable starting point for considering models of spatial
variation in economic development within regions. Logically, such analysis
has major locational implications, in that, for example, a growth impetus
transmitted between linked manufacturing industries must involve interaction in space, even if only over a short distance. It is therefore remarkable
that, to the best of the writer's knowledge, very few attempts indeed have
been made to analyse these implicatons, and in particular to formulate any
kind of model of the intra-regional locational impact of the multiplier process.
As a result, the simple locational model put forward by Tattersall (1962),
extending some of North's ideas (1955, pp. 250–251), appears as a pioneering
study in a field in which much more work needs to be done. This model,
aimed specifically at regions whose early economic growth has been geared
to exporting, usually of primary products, suggests that although the actual

production of such commodities is usually dispersed throughout different parts or the whole of the region, shipment and organization of exports soon become concentrated in certain urban centres. These centres usually grow up 'at points (frequently at natural breaks in transport routes, e.g. ports and river junctions) where transportation costs for exports and imports are minimized' (Tattersall, 1962, p. 217), and naturally handle both collection of exports and distribution of imports. As export activity expands, the operation of the multiplier process naturally stimulates the development of locally-orientated residentiary industry (interpreted here in its widest sense as referring to the production of both goods and services) within the region. The locational impact of this development is, however, also concentrated in these urban centres, partly because of the external economies available in them, partly because of 'their strategic role as distribution centres, plus their growing importance as markets'. They therefore attract 'a very large proportion of the region's domestically oriented manufacturing, and . . . a large proportion of regional trade and service employment', and grow even more rapidly (Tattersall, 1962, p. 226). Tattersall's model, although extremely simple, is interesting both for its identification of the spatial consequence of multiplier growth, and for the way in which it dovetails into North's export base model – a feature which clearly reflects the interest of both authors in the same region, the Pacific Northwest. However, much more work on this subject, aiming in particular at the integration of multiplier analysis with location theory, does appear to be needed.

(2) *The growth pole model.* Although perhaps less clearly a model than concepts already discussed, the development in recent years of a body of ideas about spatial variation in economic prosperity within regions demands attention here. Many of these ideas, explicitly or implicitly, centre around the concept of a 'growth pole', or 'growth centre'. At its simplest level, this concept refers to the fact, pointed out by many, particularly French, writers (Milhau, 1956; Gannagé, 1962, pp. 62–65; Perroux, 1961, p. 167; Beguin, 1963; Hicks, 1961, p. 77; Nicholls, 1961) and already demonstrated on other scales, that economic development never occurs uniformly over a particular region. Rather, it tends to be concentrated in certain parts, which thereby develop as 'growth poles', expanding much more rapidly than surrounding areas. These growth poles are almost invariably regarded as urban-industrial in character (e.g. Rostow, 1964, pp. 123–124; Ruttan, 1955, p. 56), while some writers have implied that they are also usually to be found close to the centre of a particular region. This latter view is represented by Schultz's claim that 'the existing economic organization works best at or near the the center of a particular matrix of economic development' (Schultz, 1953, p. 147; see also Schultz, 1950), 'matrix' here referring to a region in which economic growth is taking place, as well as by comments in two of

Friedmann's articles (1956, p. 215 and 1963, p. 55). The map in the former of these clearly indicates a central position for a growth pole, while the latter specifically outlines an 'illustrative model' in which 'metropolitan development areas' are located 'as the principal growth poles at the core of the spatial system', surrounded by concentric zones possessing other economic characteristics (Friedmann, 1963, p. 55).

At this level, of course, the growth pole concept is little more than a descriptive device. However, other writers have extended the meaning of the term to include by implication both the centre's internal mechanism of growth, and its relationships with the surrounding region. At these levels, the concept may justifiably be termed a model, in that it singles out crucial variables in the development of spatial variation in economic prosperity within a region, and specifies how they operate. The analysis of the internal expansion mechanism of a growth pole is closely associated with the work of Perroux (1955 and 1961, Part II) and other French economists. To Perroux, the growth pole ('pôle de croissance') owes its existence to the location within it of one main 'growth industry', or as he terms it 'une industrie motrice'. The growth of this industry attracts other, linked, industries (i.e. those which provide it with inputs, or derive their inputs from it) by virtue of the external economies created in the locality; and as these industries grow under stimulus from the 'industrie motrice', the growth pole as a whole expands still further. Other agglomeration economies (Friedrich, 1929, pp. 124–131; Losch, 1954, pp. 68–78) come into play to encourage further growth, a high rate of technological change is engendered by spatial proximity and ease of communication between these industries, and psychological factors, such as the development of a 'growth mentality' amongst businessmen of the district, aid further expansion (Perroux, 1955; Balassa, 1961, pp. 151 and 197).

That this kind of mechanism does work in many growth centres is attested by various studies, as for example those by Boudeville (1957) on Brazil and Lacroix (1964) on Congo. More specifically, the role of a key industry in creating external economies for supporting activities has already been noted (North, 1955, p. 252; Tattersall, 1962, p. 217), while further evidence on this is provided by Chinitz (1961) in his study of industrial agglomeration in Pittsburgh, Harris (1959), dealing with urbanization in India, and Pred (1965), analysing the relationship between industrialization and urban growth in the United States. The psychological advantages of a growth centre are stressed by Hirschman (1958, pp. 185–186). However, although the development of agglomeration economies does provide a powerful mechanism for continued industrial and urban growth, Perroux's insistence on the role of an 'industrie motrice' cannot really be accepted, for two reasons. Firstly, as Chinitz (1961) has shown, the very presence of just such a growth industry in an urban centre may in the long run inhibit industrial and economic development, by preventing the ingress of firms in other industries. In Pittsburgh, for example, the dominance of the primary metals industry over local capital and labour markets acts as a barrier to the development of such new firms; and Pittsburgh's recent overall economic growth has almost certainly suffered because of this. Secondly, and more important, many obvious growth poles exist which are not dominated by an 'industrie motrice'. For instance, Burley (1962, p. 184) rightly describes Mexico City as 'an outstanding example' of recent concentration of industrial and economic development (see also Bird, 1963); yet no one growth industry can be identified as stimulating this process, a very wide range of unconnected activities having grown up here over the last forty years. The same is true of other growth centres, such as those of Jinja, Uganda (Hoyle, 1963 and 1964) and Nairobi, Kenya (Pollock, 1960, p. 352), or the booming metropolis of São Paulo, Brazil (James, 1959, p. 500). For these reasons, Perroux's analysis of the role of an 'industrie motrice' in growth pole development, though valid in some cases, and probably of value for planning government-sponsored growth centres, cannot be accepted as the typical mechanism of internal expansion of a growth pole.

The third important feature of the growth pole model is its analysis of relationships between the growth centre and surrounding region – an analysis which, together with that of agglomeration economies, largely explains the recent acceptance of the model by several governments as a key concept in regional economic planning (Harris, 1966, p. 577). Probably the earliest example of this acceptance is represented by the Brazilian government's decision to locate the country's new capital, Brasilia, in the undeveloped interior of Brazil, 'the most often heard and perhaps the most persuasive single argument' for this being 'the profound reciprocal effect' which the city was intended to exercise on the economy of its surrounding region (Snyder, 1964, p. 35). However, other countries, notably France (Fourastie and Courthéoux, 1963, p. 134), Italy (Coquery, 1964) and the United Kingdom (National Economic Development Council, 1963, pp. 14–29; Board of Trade, 1963, p. 6; Scottish Development Department, 1963, pp. 27–30) have in recent years also apparently come to regard the model as providing a basis for the practical planning of regional development (Parr, 1965, pp. 1, 5). Indeed, in the United Kingdom if not elsewhere, certain new towns are now being planned deliberately as growth centres (Diamond, 1965, p. 183), in contrast to earlier new town policy (Rodwin, 1955, pp. A2–A3). This general acceptance partly reflects the belief, derived by logical extension from the model's analysis of agglomeration economies, that concentration of government investment and industrial activity within a growth pole will, in the long run, stimulate a higher level of industrial and economic development than if it had been spread over a wider region (see, for example, Scottish Development Department, 1963, p. 27). However, it also clearly reflects the idea that the 'benefit of new growth in any part would repercuss fairly quickly throughout the region' (National Economic Development Council, 1963, p. 26), or, in Wright's phrase (1965, p. 150), 'the hope that prosperity will spread outwards from the chosen points in concentric ripples'. If this prospect were not in fact suggested by the model, governments would undoubtedly have been far more chary of basing their regional development programmes on it.

This idea of centrifugal spread of economic prosperity from a growth pole probably owes something to earlier views on the functional relationship between a large city and its surrounding hinterland, such as those of Gras (1922, p. 700). However, the first explicit discussion of the relationship between urban economic growth and that of the surrounding region came in a group of articles published in the early 1950's. In the earliest of these, Hoselitz (1953) examined the role of medieval European cities in regional economic growth, and concluded that the 'increase in average real income' in these cities produced by industrialization and the development of efficient governmental services 'also strongly affected the non-urban regions located near the centers of development' (Hoselitz, 1953, p. 203). One important way in which this occurred was through a growth in urban demand for labour, and the development of commuting from villages close to the towns. This analysis was extended in his later study (Hoselitz, 1955B), which classified cities as 'generative' or 'parasitic' on the basis of whether or not they stimulated the economic growth of the wider region in which they were located. Many of the early colonial settlements in the New World and South Africa, Hoselitz claimed, were parasitic, enjoying a certain degree of economic growth 'within the city itself and its surrounding environs' only at the expense of the rest of the region, which was ruthlessly exploited for its natural and agricultural resources (Hoselitz, 1955B, p. 280). Overall regional growth was thereby retarded. However, Hoselitz does not appear to regard this as the normal regional impact of urban development in underdeveloped countries, since he points out that even these parasitic colonial settlements eventually developed into generative centres. Such centres transmit a growth stimulus to the surrounding area in various ways. In addition to the labour aspect already mentioned, economic and industrial development in generative cities (a) creates a new demand for industrial raw materials from the surrounding region, and (b) attracts new population to the cities, thereby increasing the demand for food from the countryside. The net effect of these forces is a 'widening of economic development over an increasing area affecting a growing proportion of the population outside the city' (Hoselitz, 1955B, p. 282).

Other writers appear to support Hoselitz's conclusions. For example, both Lampard (1954-55) and Stolper (1954-55) agree that large cities in underdeveloped countries have sometimes been parasitic, acting as 'a curb rather than a stimulus to wider economic growth' (Lampard, 1954-55, p. 131). Both writers suggest that a key factor in this is the dissipation of wealth derived from the surrounding region in non-productive urban consumption. Rather than investment in industry, for example, such wealth is used for 'grandiose urban construction' (Lampard, 1954–55, p. 131), which stimulates little or no economic development in the wider region (see also Tangri, 1962, p. 209). Cultural and racial factors may also act as major barriers to growth transmission (Stolper, 1954–55; Deane, 1961, p. 19; Hicks, 1961, p. 77). On the other hand, where these barriers are weaker, and urban investment is channeled into productive enterprises, urban economic development in an underdeveloped country does apparently stimulate growth in the surrounding region. Such at least is claimed by Myrdal (1957B, p. 31) and Friedmann (1961, pp. 95–96), and seems to have occurred, for example, in Uganda where 'the process of economic development which changed all of southern Uganda' during the colonial period 'radiated from the Protectorate center of commerce and colonial policy, Kampala in the Kingdom of Buganda' (Larimore, 1960, p. 120).

Hoselitz's ideas on this topic are more developed than those of other writers such as Myrdal (1957B, p. 31). However, they are clearly aimed primarily at underdeveloped countries, and, as has been pointed out (Lombardini, 1964; Kindleberger, 1964, p. 263), regional development in an underdeveloped area is a different matter from that in the more advanced economies. In the latter, whether the development problem is posed by a distressed industrial region (Estall, 1964) or a growing metropolitan region (Wright, 1965, pp. 148–149), economic and social conditions (such as much higher levels of car ownership) would appear to be even more conducive to the transmission of economic prosperity from a growth pole to its surrounding region.

This has been explicitly recognized by several commentators. Gerschenkron (1963), for example, has pointed out that in the United States 'the economic difference between the city and the countryside is no longer what it used to be' (Gerschenkron, 1963, p. 59) – i.e. economic prosperity has spread from the towns to surrounding areas. Key factors in this spread have been growing decentralization of industry, housing and even shops from the towns to surrounding areas, together with increasing car and refrigerator ownership, the latter permitting a considerably enlarged spatial scale of journeys-to-work and journeys-to-shop. Though Gerschenkron himself does not mention this, increased recreational use of surrounding rural areas by urban dwellers represents another mechanism by which prosperity has been spread within regions of the United States. Interestingly enough, the value of a simple growth pole model in planning resultant recreational development has recently been stressed by Harper, Schmudde and Thomas (1966). Thus intra-regional spread of prosperity is a direct function of the already high level of economic development in the United States, or, as Gerschenkron (1963, p. 61) puts it, of 'the rise in incomes and technological progress'. Similarly,

5

Wright (1965, p. 161) has stressed that in Britain 'the best position now for the go-ahead family and firm is a position near but not in a big city', a point clearly illustrated by both recent industrial and residential decentralization from major growth poles such as Greater London (Keeble, 1965, pp. 24–28; Pahl, 1965). Again, an intra-regional spread of economic prosperity is the result. However, the only commentator to develop these ideas into a growth pole model is Martin (1957). Adapting Schultz's views, Martin (1957, p. 173) describes 'a model of a dynamic, industrializing society in which economic development occurs primarily in the urban type of locational matrix'. This development in time produces a spread of prosperity to surrounding areas (a) by tapping them for industrial and urban raw materials and customers, (b) by stimulating food production for urban markets, and the introduction to the countryside of industrial-type farming techniques, (c) by encouraging migration of surplus rural population to the town, and (d) by deconcentration of urban population, industry and other institutions (Martin, 1957, pp. 173–174). Like Gerschenkron, Martin stresses that the fourth kind of impact is dependent upon a very high existing level of economic development.

Very little empirical testing of this kind of growth model has as yet been carried out. However, two studies of the changing pattern of agricultural development and farm income in the Tennessee Valley do throw limited light on its relevance to that area. In the earlier, Ruttan (1955) demonstrated by regression analysis that median farm family incomes in different parts of the Valley were closely correlated with the level of urban-industrial development in the same areas. In other words, proximity to urban-industrial centres had stimulated agricultural prosperity in some areas, chiefly through provision of increased off-farm jobs for members of farm families, but also through greater availability of local capital and markets. On the whole, this finding supports Martin's model. Nicholl's later study (1961) also found a close connection between proximity to urban centres and agricultural prosperity: but in stressing the increasing divergence over the 1900–50 period of intra-regional indexes of farm prosperity, his analysis implies that spread from urban growth poles has only influenced areas within a relatively short distance of these centres. Clearly, scale of regional size is very important when assessing the effectiveness of intra-regional spread of economic development from a growth pole.

The limitations of these two studies clearly illuminate the need for much more empirical testing and refining of the growth pole model – whether applied to underdeveloped or developed regions. Indeed, in view of its acceptance by governments in advanced economies as an important concept in regional planning, it is remarkable that so little work has been done on it, either in terms of spatial interaction between pole and region, or of internal growth mechanisms. What limited analysis has been carried out, as for example by Humphrys (1965) on the role of service industry in the expansion of growth centres, tends to throw up as many questions as it answers. Here surely is an important field for future geographical work.

CONCLUSION

Particularly at national and sub-national scales, then, the building of models of economic development is now proceeding rapidly, stimulated by a growing realization on the part of economists and government planners of the value of this approach for practical and theoretical purposes. Geographers have so far played very little part in this, despite their genuine interest in the variation over the earth's surface of many of the phenomena which contribute to spatial variations in economic development. In the light of a changing internal and external intellectual environment (Wrigley, 1965, pp. 13–19), however, geography surely now needs to accept model-building as an important avenue of study, and, in particular, to formulate and refine models of spatial variation and interaction in economic development.

REFERENCES

ACKERMAN, E. A., [1958], Geography as a Fundamental Research Discipline; *University of Chicago, Department of Geography Research Paper*, 53, 37 pp.
ACKERMAN, E. A., [1963], Where is a Research Frontier?; *Annals of the Association of American Geographers*, 53 (4), 429–440.
ALEXANDER, J. W., [1954], The Basic-Nonbasic Concept of Urban Economic Functions; *Economic Geography*, 30 (3), 246–261.
ANDIC, S. and PEACOCK, A. T., [1961], The International Distribution of Income, 1949 and 1957; *Journal of the Royal Statistical Society, Series A*, 124 (2), 206–218.
ANDREWS, R. B., [1953–56], Mechanics of the Urban Economic Base; *Land Economics*, 29–31.
BACHMURA, F. T., [1959], Man-Land Equalization through Migration; *American Economic Review*, 49 (5), 1004–1017.
BAER, W., [1964], Regional Inequality and Economic Growth in Brazil; *Economic Development and Cultural Change*, 12 (3), 268–285.
BALASSA, B., [1961], *The Theory of Economic Integration*, (Homewood, Ill.), 304 pp.
BALDWIN, R. E., [1956], Patterns of Development in Newly Settled Regions; *Manchester School of Economic and Social Studies*, 24 (2), 161–179.
BARAN, P. A. and HOBSBAWM, E. J., [1961], The Stages of Economic Growth; *Kyklos*, 14, 324–342.
BARFOD, B., [1938], *Local Economic Effects of A Large-scale Industrial Undertaking*, (Copenhagen), 74 pp.
BARNA, T., (Ed.), [1963], *Structural Interdependence and Economic Development*, (London), 365 pp.

BARZANTI, S., [1965], *Underdeveloped Areas within the Common Market*, (Princeton), 456 pp.

BEGUIN, H., [1963], Aspects Géographiques de la Polarisation; *Tiers-Monde*, 4, 16, 559–608.

BERMAN, E. B., [1959], A Spatial and Dynamic Growth Model; *Papers & Proceedings of the Regional Science Association*, 5, 143–150.

BERRY, B. J. L., [1960], An Inductive Approach to the Regionalization of Economic Development; Ch. 6, in Ginsburg, N., (Ed.), Essays on Geography and Economic Development; *University of Chicago, Department of Geography Research Paper*, 62, 173 pp.

BERRY, B. J. L., [1961A], City Size Distributions and Economic Development; *Economic Development and Cultural Change*, 9 (4), Part I, 573–587.

BERRY, B. J. L., [1961B], Basic Patterns of Economic Development; Part VIII, In Ginsburg, N., *Atlas of Economic Development*, (Chicago), 110–119.

BIRD, R., [1963], The Economy of the Mexican Federal District; *Inter-American Economic Affairs*, 17 (2), 19–51.

BOARD OF TRADE, [1963], *The North East. A Programme for Regional Development and Growth*, (London), 48 pp.

BORTS, G. H., [1960], The Equalization of Returns and Regional Economic Growth; *American Economic Review*, 50 (3), 319–347.

BORTS, G. H. and STEIN, J. L., [1964], *Economic Growth in a Free Market*, (New York), 235 pp.

BOUDEVILLE, J. R., [1957], Contribution a l'étude des pôles de croissance brésiliens; *Cahiers de l'Institute de Science Economique Appliquée, Cahiers disponibles, Série F*, 10, 71 pp.

BRANDENBURG, F. R., [1964], *The Making of Modern Mexico*, (Englewood Cliffs), 379 pp.

BRIGGS, A., [1965], Technology and Economic Development; In *Technology and Economic Development*, (Harmondsworth, Middlesex), 15–32.

BUNGE, W., [1962], Theoretical Geography; *Lund Studies in Geography, Series C, General and Mathematical Geography*, 1, 201 pp.

BURLEY, T. M., [1962], Industrial Expansion in the Federal District, Mexico; *Geography*, 47 (2), 184–185.

CAESAR, A. A. L., [1964], Planning and the Geography of Great Britain; *Advancement of Science*, 21, 91, 230–240.

CAIRNCROSS, A. K., [1959], Research on Comparative Economic Growth; In National Bureau of Economic Research, *The Comparative Study of Economic Growth and Structure*, (New York), 106–109.

CAIRNCROSS, A. K., [1961], Essays in Bibliography and Criticism, XLV, The Stages of Economic Growth; *Economic History Review, Second Series*, 13 (3), 450–458.

CAIRNCROSS, A. K., [1962], *Factors in Economic Development*, (London), 346 pp.

CAMERON, G. C. and CLARK, B. D., [1966], Industrial Movement and the Regional Problem; *University of Glasgow Social & Economic Studies, Occasional Papers*, 5, 220 pp.

CAVES, R. E., [1960], *Trade and Economic Structure: Models and Methods*, (Cambridge, Mass.), 317 pp.

CHATTERJI, M. K., [1964], An Input-Output Study of the Calcutta Industrial Region; *Papers of the Regional Science Assocation*, 13, 93–102.

CHENERY, H. B., [1956], Inter-Regional and International Input-Output Analysis; In Barna, T., (Ed.), *The Structural Interdependence of the Economy*, (New York), 339–356.

CHENERY, H. B., [1960], Patterns of Industrial Growth; *American Economic Review*, 50 (4), 624–654.

CHENERY, H. B., [1961], Comparative Advantage and Development Policy; *American Economic Review*, 51 (1), 18–51.

CHENERY, H. B., [1962], Development Policies for Southern Italy; *Quarterly Journal of Economics*, 76 (4), 515–547.

CHENERY, H. B., [1963], The Use of Interindustry Analysis in Development Programming; In Barna, T., (Ed.), *Structural Interdependence and Economic Development*, (London), Ch. 1.

CHENERY, H. B., CLARK, P. G. and CAO-PINNA, V., [1953], *The Structure and Growth of the Italian Economy*, (Rome), 165 pp.

CHENERY, H. B. and CLARK, P. G., [1959], *Interindustry Economics*, (New York), 345 p.

CHINITZ, B., [1961], Contrasts in Agglomeration: New York and Pittsburgh; *American Economic Review*, 51 (2), 279–289.

CHISHOLM, M., [1962], Tendencies in Agricultural Specialization and Regional Concentration of Industry; *Papers of the Regional Science Association, European Congress*, 10, 157–162.

CHISHOLM, M., [1966], *Geography and Economics*, (London), 230 pp.

CHORLEY, R. J., [1964], Geography and Analogue Theory; *Annals of the Association of American Geographers*, 54 (1), 127–137.

CHORLEY, R. J. and HAGGETT, P., (Eds.), [1965], *Frontiers in Geographical Teaching*, (London), 378 pp.

CLARK, C., [1966], Industrial Location and Economic Potential; *Lloyds Bank Review*, 82, 1–17.

CLAWSON, M. (Ed.), [1964], *Natural Resources and International Development*, (Baltimore), 462 pp.

CLOUGH, S. B. and LIVI, C., [1956], Economic Growth in Italy: An Analysis of the Uneven Development of North and South; *Journal of Economic History*, 16 (3), 334–349.

COHEN, K. J. and CYERT, R. M., [1961], Computer Models in Dynamic Economics; *Quarterly Journal of Economics*, 75 (1), 112–127.

COMMITTEE FOR ECONOMIC DEVELOPMENT, [1958], *The 'Little Economies': Problems of United States Area Development*, (New York), 60 pp.

CONRAD, A. H. and MEYER, J. R., [1965], *Studies in Economic History*, (London), 241 pp.

COQUERY, M., [1964], Problèmes de Développement et Aspects Nouveaux de L'Industrialisation en Italie Méridionale; *Bulletin de la Section de Géographie, Ministère de L'Education Nationale, Comité de Travaux Historiques et Scientifiques*, 76, Etudes Méditerranéenes, 393—498.

COUTSOUMARIS, G., [1964], Regional Activity Relocation Problems in a Developing

Economy; *Papers of the Regional Science Association, European Congress*, 12, 79–86.

DALY, M. C., [1940], An Approximation to a Geographical Multiplier; *Economic Journal*, 50, 248–258.

DAS-GUPTA, A. K., [1965], *Planning and Economic Growth*, (London), 185 pp.

DEANE, P., [1961], The Long Term Trends in World Economic Growth; *Malayan Economic Review*, 6 (2), 14–26.

DEPARTMENT OF APPLIED ECONOMICS, University of Cambridge, [1962], A Computable Model of Economic Growth; *A Programme for Growth*, 1, (London), 91 pp.

DIAMOND, D. R., [1965], Regional Planning: The Scottish Approach; pp. 183–184, in Caesar, A. A. L. and Keeble, D. E., (Eds.), Regional Planning Problems in Great Britain; *Advancement of Science*, 22, 97, 177–185.

DRUMMOND, I., [1961], Review of 'The Stages of Economic Growth' by W. W. Rostow; *Canadian Journal of Economics and Political Science*, 27 (1), 112–113.

DUESENBERRY, J. S., [1950], Some Aspects of the Theory of Economic Development; *Explorations in Entrepreneurial History*, 3 (2), 63–102.

DWYER, D. J., [1965], Size as a Factor in Economic Growth: Some Reflections on the Case of Hong Kong; *Tijdschrift voor Economische en Sociale Geografie*, 56 (5), 186–192.

EASTERLIN, R. A., [1958], Long Term Regional Income Changes: Some Suggested Factors; *Papers & Proceedings of the Regional Science Association*, 4, 313–325.

EASTERLIN, R. A., [1960], Interregional Differences in Per Capita Income, Population, and Total Income, 1840–1950; In National Bureau of Economic Research, Conference on Research in Income and Wealth, *Trends in the American Economy in the Nineteenth Century*, (Princeton), 73–140.

ECKAUS, R. S., [1961], The North–South Differential in Italian Economic Development; *Journal of Economic History*, 21 (3), 285–317.

ECONOMIC COMMISSION FOR EUROPE, Research and Planning Division, [1955], Problems of Regional Development and Industrial Location in Europe; In *Economic Survey of Europe in 1954*, (Geneva), 136–171.

EL-KAMMASH, M. M., [1963], On the Measurement of Economic Development using Scalogram Analysis; *Papers & Proceedings of the Regional Science Association*, 11, 309–334.

ELKAN, W., [1959], Regional Disparities in the Incidence of Taxation in Uganda; *Review of Economic Studies*, 26, 70, 135–143.

ENKE, S., [1964], *Economics for Development*, (London), 616 pp.

ESTALL, R. C., [1964], Planning for Industry in the Distressed Areas of the U.S.; *Journal of the Town Planning Institute*, 50 (9), 390–396.

EUROPEAN COAL AND STEEL COMMUNITY, HIGH AUTHORITY, [1961], *Les Politiques Nationales de Développement Régional et Conversion*, (Brussels), 196 pp.

EUROPEAN FREE TRADE ASSOCIATION, [1965], *Regional Development Policies in EFTA*, (Geneva), 78 pp.

EWING, A. F., [1964], Industrialisation and the U.N. Economic Commission for Africa; *Journal of Modern African Studies*, 2 (3), 351–363.

FAIRBANK, J. K., ECKSTEIN, A. and YANG, L. S., [1960], Economic Change in

Early Modern China: An Analytic Framework; *Economic Development and Cultural Change*, 9 (1), 1–26.

FISHER, J. L., [1955], Concepts in Regional Economic Development; *Papers & Proceedings of the Regional Science Association*, 1, W1–W20.

FISHER, M. R., [1964], Macro-Economic Models: Nature, Purpose and Limitations; *Eaton Papers*, 2, 40 pp.

FLORENCE, P. S., [1948], Investment, Location, and Size of Plant; *National Institute of Economic and Social Research, Economic and Social Studies*, 7, 211 pp.

FORDHAM, P., [1965], Natural Resources and Economic Development; In *The Geography of African Affairs*, (Harmondsworth), Ch. 4.

FOURASTIE, J. and COURTHÉOUX, J-P., [1963], *La Planification Economique en France*, (Paris), 208 pp.

FRIEDMANN, J. R. P., [1956], Locational Aspects of Economic Development; *Land Economics*, 32 (3), 213–227.

FRIEDMANN, J. R. P., [1959], Regional Planning: A Problem in Spatial Integration; *Papers & Proceedings of the Regional Science Association*, 5, 167–179.

FRIEDMANN, J. R. P., [1961], Integration of the Social System: An Approach to the Study of Economic Growth; *Diogenes*, 33, 75–97.

FRIEDMANN, J. R. P., [1963], Regional Economic Policy for Developing Areas; *Papers & Proceedings of the Regional Science Association*, 11, 41–61.

FRIEDMANN, J. R. P. and ALONSO, W. (Eds.), [1964], *Regional Development and Planning: A Reader*, (Cambridge, Mass.), 722 pp.

FRIEDRICH, C. J., [1929], *Alfred Weber's Theory of the Location of Industries*, (Chicago), 256 pp.

FRYER, D. W., [1958], World Income and Types of Economies: The Pattern of World Economic Development; *Economic Geography*, 34 (4), 283–303.

FRYER, D. W., [1965], *World Economic Development*, (New York), 627 pp.

FULMER, J. L., [1950], Factors Influencing State Per Capita Income Differentials; *Southern Economic Journal*, 16 (3), 259–278.

FURTADO, C., [1963], *The Economic Growth of Brazil. A Survey from Colonial to Modern Times*, (Berkeley), 285 pp.

FURTADO, C., [1964], *Regional Development in Brazil*, (Berkeley).

FYOT, J-L. and CALVEZ, J-Y., [1956], *Politique Economique Régionale en Grande-Bretagne*, (Paris), 312 pp.

GANNAGÉ, E., [1962], *Economie du Développement*, (Paris), 356 pp.

GERSCHENKRON, A., [1962], *Economic Backwardness in Historical Perspective*, (Cambridge, Mass.), 456 pp.

GERSCHENKRON, A., [1963], City Economies – Then and Now; In Handlin O. and Burchard, J., (Eds.), *The Historian and the City*, (Cambridge, Mass.), 56–62.

GILMORE, D. R., [1960], *Developing the Little Economies*; *A Survey of Area Development Programs in the United States*, (New York), 200 pp.

GINSBURG, L. B., [1957], Regional Planning in Europe; *Journal of the Town Planning Institute*, 43 (6), 142–147.

GINSBURG, N., [1957], Natural Resources and Economic Development; *Annals of the Association of American Geographers*, 47 (3), 197–212.

GINSBURG, N. (Ed.), [1960], Essays on Geography and Economic Development; *University of Chicago, Department of Geography Research Paper* 62, 173 pp.

GINSBURG, N., [1961], *Atlas of Economic Development*, (Chicago), 119 pp.

GLADE, W. P. and ANDERSON, C. W., [1963], *The Political Economy of Mexico*, (Madison, Wisconsin), 242 pp.

GOLDSMITH, R. W., [1959], Explanatory Report; In National Bureau of Economic Research, *The Comparative Study of Economic Growth and Structure*, (New York), Part I, 3–100.

GRAS, N. S. B., [1922], The Development of Metropolitan Economy in Europe and America; *American Historical Review*, 17 (4), 695–708.

GRAS, N. S. B., [1930], Stages in Economic History; *Journal of Economic and Business History*, 2, 397.

GREEN, L. P. and FAIR, T. J. D., [1962], *Development in Africa*, (Johannesburg), 203 pp.

GRIBAUDI, F., [1965], Some Geographic Aspects of Economic Development; *Tijdschrift voor Economische en Sociale Geografie*, 56 (2), 69–72.

GUTHRIE, J. A., [1955], Economies of Scale and Regional Development; *Papers & Proceedings of the Regional Science Association*, 1, J1–J10.

HABAKKUK, J. H., [1961], Review of The Stages of Economic Growth by W. W. Rostow; *Economic Journal*, 71, 283, 601–604.

HABERLER, G., [1959], *International Trade and Economic Development*, (Cairo), 36 pp.

HAGEN, E. E., [1959], Economic Structure and Economic Growth: A Survey of Areas in which Research is Needed; In National Bureau of Economic Research, *The Comparative Study of Economic Growth and Structure*, (New York), 124–141.

HAGEN, E. E., [1960], Some Facts about Income Levels and Economic Growth; *Review of Economics and Statistics*, 42 (1), 62–67.

HAGEN, E. E., [1962], *On the Theory of Social Change: How Economic Growth Begins*, (Cambridge, Mass.), 557 pp.

HAGGETT, P., [1965A], Changing Concepts in Economic Geography; Ch. 6 in Chorley, R. J. and Haggett, P. (Eds.), *Frontiers in Geographical Teaching*, (London), 378 pp.

HAGGETT, P., [1965B], Scale Components in Geographical Problems; Ch. 9 in Chorley, R. J. and Haggett, P. (Eds.), *Frontiers in Geographical Teaching*, (London), 378 pp.

HAGGETT, P., CHORLEY, R. J. and STODDART, D. R., [1965], Scale Standards in Geographical Research: A New Measure of Areal Magnitude; *Nature*, 205 4974, 844–847.

HANCE, W. A., [1958], *African Economic Development*, (London), 307 pp.

HAQ, M. U., [1963], *The Strategy of Economic Planning*, (Karachi), 266 pp.

HARPER, R. A., SCHMUDDE, T. H. and THOMAS, F. H., [1966], Recreation Based Economic Development and the Growth-Point Concept; *Land Economics*, 42 (1), 95–101.

HARRIS, B., [1959], Urbanisation Policy in India; *Papers & Proceedings of the Regional Science Association*, 5, 181–203.

HARRIS, D., [1966], The Idea of the Growth Area; *Official Architecture and Planning*, 29 (4), 577–581.

HARRIS, S. E., [1954], Interregional Competition: With Particular Reference to North–South Competition; *American Economic Review*, 44 (2), 367–380.

HARRIS, S. E., [1957], *International and Interregional Economics*, (New York), 564 pp.

HART, P. E., MILLS, G. and WHITAKER, J. K. (Eds.), [1964], *Econometric Analysis for National Economic Planning*, (London), 320 pp.

HARTSHORNE, R., [1939], *The Nature of Geography*, (Lancaster, Penn.), 482 pp.

HARTSHORNE, R., [1959], *Perspective on the Nature of Geography*, (Chicago), 201 pp.

HATHAWAY, D. E., [1960], Migration from Agriculture: The Historical Record and Its Meaning; *American Economic Review*, 50 (2), 379–391.

HAY, A. M. and SMITH, R. H. T., [1966], Preliminary Estimates of Nigeria's Interregional Trade and Associated Money Flows; *Nigerian Journal of Economic and Social Studies*, 8 (1), 9–35.

HEMMING, M. F. W., [1963], The Regional Problem; *National Institute Economic Review*, 25, 40–57.

HICKS, J. R., [1953], Review of The Process of Economic Growth by W. W. Rostow; *Journal of Political Economy*, 61 (2), 173–174.

HICKS, J. R., [1959], *Essays in World Economics*, (Oxford), 274 pp.

HICKS, U. K., CARNELL, F. G., NEWLYN, W. T., HICKS, J. R. and BIRCH, A. H., [1961], *Federalism and Economic Growth in Underdeveloped Countries*, (London), 185 pp.

HIGGINS, B., [1964], Review of On the Theory of Social Change: How Economic Growth Begins, by E. E. Hagen; *Journal of Political Economy*, 72 (6), 627–630.

HILDEBRAND, G. H. and MACE, A., [1950], The Employment Multiplier in an Expanding Industrial Market: Los Angeles County, 1940–47; *Review of Economics and Statistics*, 32 (3), 241–249.

HIRSCH, W. Z., [1959], Interindustry Relations of a Metropolitan Area; *Review of Economics and Statistics*, 41 (4), 360–369.

HIRSCH, W. Z., [1962], Design and Use of Regional Accounts; *American Economic Review*, 52 (2), 365–373.

HIRSCH, W. Z., [1963], Application of Input-Output Techniques to Urban Areas; Ch. 8 in Barna, T. (Ed.), *Structural Interdependence and Economic Development*, (Geneva), 365 pp.

HIRSCHMAN, A. O., [1958], *The Strategy of Economic Development*, (New Haven, Connecticut), 217 pp.

HIRSCHMAN, A. O., [1959], Some Suggestions for Research on Comparative Development; In National Bureau of Economic Research, *The Comparative Study of Economic Growth and Structure*, (New York), 142–144.

HOCHWALD, W., (Ed.), [1961], *The Design of Regional Accounts*, (Baltimore), 281 pp.

HOOVER, E. M., [1937], *Location Theory and the Shoe and Leather Industries*, (Cambridge, Mass.), 323 pp.

HOOVER, E. M., [1948], *The Location of Economic Activity*, (New York), 310 pp.

HOSELITZ, B. F., [1953], The Role of Cities in the Economic Growth of Underdeveloped Countries; *Journal of Political Economy*, 61 (3), 195–208.

HOSELITZ, B. F., [1955A], Patterns of Economic Growth; *Canadian Journal of Economics & Political Science*, 21 (4), 416–431.

HOSELITZ, B. F., [1955B], Generative and Parasitic Cities; *Economic Development and Cultural Change*, 3, 278–294.

HOSELITZ, B. F., [1959], On Historical Comparisons in the Study of Economic Growth; In National Bureau of Economic Research, *The Comparative Study of Economic Growth and Structure*, (New York), 145–161.

HOSELITZ, B. F., [1960], *Theories of Economic Growth*, (New York), 344 pp.

HOUGHTON, D. H., [1964], *The South African Economy*, (Cape Town), 261 pp.

HOYLE, B. S., [1963], The Economic Expansion of Jinja, Uganda; *Geographical Review*, 53 (3), 377–388.

HOYLE, B. S., [1964], Further Industrial Growth at Jinja; *East African Geographical Review*, 2, 44–45.

HUGHES, R. B., [1961], Interregional Income Differences; Self-Perpetuation; *Southern Economics Journal*, 28 (1), 41–45.

HUMPHRYS, G., [1962], Growth Industries and the Regional Economies of Britain; *District Bank Review*, 144, 35–56.

HUMPHRYS, G., [1963], Governmental Policy and the Growth Industries; *Professional Geographer*, 15 (4), 13–16.

HUMPHRYS, G., [1965], Services in Growth Centres, and their Implications for Regional Planning, with Special Reference to South Wales; In Caesar, A. A. L. and Keeble, D. E., (Eds.), Regional Planning Problems in Great Britain; *Advancement of Science*, 22, 181–182.

INTERNATIONAL INFORMATION CENTRE FOR LOCAL CREDIT, [1964], *Government Measures for the Promotion of Regional Economic Development*, (The Hague), 159 pp.

ISARD, W., [1951], Interregional and Regional Input-Output Analysis: A Model of a Space-Economy; *Review of Economics and Statistics*, 33 (4), 318–328.

ISARD, W., [1958], Interregional Linear Programming: An Elementary Presentation and a General Model; *Journal of Regional Science*, 1 (1), 1–59.

ISARD, W., [1960], *Methods of Regional Analysis: an Introduction to Regional Science*, (New York), 784 pp.

ISARD, W. and CUMBERLAND, J. H., (Eds.), [1961], *Regional Economic Planning, Techniques of Analysis for Less Developed Areas*, (Paris), 450 pp.

ISARD, W. and KUENNE, R., [1953], The Impact of Steel upon the Greater New York-Philadelphia Industrial Region: A Study in Agglomeration Projection; *Review of Economics and Statistics*, 35 (4), 289–301.

ISARD, W. and REINER, T., [1961], Regional and National Economic Planning and Analytic Techniques for Implementation; In Isard, W. and Cumberland, J. H., (Eds.), *Regional Economic Planning, Techniques of Analysis for Less Developed Areas*, (Paris), 19–38.

ISARD, W. and SMOLENSKY, E., [1963], Application of Input-Output Techniques to Regional Science; Ch. 6, in Barna, T. (Ed.), *Structural Interdependence and Economic Development*, (London), 365 pp.

JAMES, P. E., [1951], An Assessment of the Role of the Habitat as a Factor in Differential Economic Development; *American Economic Review*, 41 (2), 229–238.

JAMES, P. E., [1959], *Latin America*, (London), 942 pp.

JOHNSTON, B. F. and MELLOR, J. W., [1961], The Role of Agriculture in Economic Development; *American Economic Review*, 51 (4), 566–593.

JOUANDET-BERNADAT, R., [1964], Les comptabilités économiques régionales; *Revue D'Economie Politique*, 74 (1), 136–168.

KEEBLE, D. E., [1965], Industrial Migration from North-West London, 1940–1964; *Urban Studies*, 2 (1), 15–32.

KEIRSTEAD, B. S., [1948], *The Theory of Economic Change*, (Toronto), 386 pp.

KELLER, F., [1953], Resources Inventory – A Basic Step in Economic Development; *Economic Geography*, 29 (1), 39–47.

KINDLEBERGER, C. P., [1958], *Economic Development*, (New York), 325 pp.

KINDLEBERGER, C. P., [1964], *Economic Growth in France and Britain, 1851–1950*, (Cambridge, Mass.), 378 pp.

KLAASSEN, L. H., [1965], *Area Economic and Social Redevelopment*, (Paris), 113 pp.

KLAASSEN, L. H., KROFT, W. C. and VOSKUIL, R., [1963], Regional Income Differences in Holland; *Papers of the Regional Science Association, European Congress*, 10, 77–81.

KLEIN, L. R., [1961], A Model of Japanese Economic Growth; *Econometrica*, 29 (3), 277–292.

KUENNE, R. E., [1955], A Rejoinder; *Review of Economics and Statistics*, 37 (3), 312–314.

KUZNETS, S., [1951], The State as a Unit in Study of Economic Growth; *Journal of Economic History*, 11 (1), 25–41.

KUZNETS, S., [1953–54], International Differences in Income Levels: Reflections on their Causes; *Economic Development and Cultural Change*, 2, 3–26.

KUZNETS, S., [1955], Economic Growth and Income Inequality; *American Economic Review*, 45 (1), 1–28.

KUZNETS, S., [1956], Quantitative Aspects of the Economic Growth of Nations, 1 – Levels and Variability of Rates of Growth; *Economic Development and Cultural Change*, 5, 5–94.

KUZNETS, S., [1958], Economic Growth of Small Nations; In Bonne, A., (Ed.), *The Challenge of Development*, (Jerusalem), 9–25.

KUZNETS, S., [1959], On Comparative Study of Economic Structure and Growth of Nations; In National Bureau of Economic Research, *The Comparative Study of Economic Growth and Structure*, (New York), 162–176.

KUZNETS S., [1960], Economic Growth of Small Nations; Ch. 2 in Robinson, E. A. G., (Ed.), *Economic Consequences of the Size of Nations*, (London), 447 pp.

KUZNETS, S., [1963], Notes on the Take-Off; Ch. 2 in Rostow, W. W., (Ed.), *The Economics of Take-Off into Sustained Growth*, (London), 482 pp.

LACOSTE, Y., [1962], Le sous-développement: quelques ouvrages significatifs parus depuis dix ans; *Annales de Geographie*, 71, 385–386, 247–278 and 387–414.

LACROIX, J. L., [1964], Les Poles de Développement Industrial en Congo; *Cahiers Economiques et Sociaux, Institute de Recherches Economiques et Sociales, Université Lovanium*, 11 (1), 146–191.

LAMPARD, E. E., [1954–55], The History of Cities in the Economically Advanced Areas; *Economic Development and Cultural Change*, 3, 81–136.

LANE, T., [1966], The Urban Base Multiplier: An Evaluation of the State of the Art; *Land Economics*, 42 (3), 339–347.

LARIMORE, A., [1960], A Measure of Economic Change: Sequent Development of Occupance in Busoga District, Uganda; In Ginsburg, N., (Ed.), Essays on Geography and Economic Development; *University of Chicago, Department of Geography Research Paper* 62, 111–123.

LASUEN, J. R., [1962], Regional Income Inequalities and the Problems of Growth in Spain; *Papers of the Regional Science Association, European Congress*, 8, 169–191.

LEONTIEF, W. W., [1936], Quantitative Input-Output Relations in the Economic System of the United States; *Review of Economics and Statistics*, 18 (3), 103–125.
LEONTIEF, W. W., [1951A], Input-Output Economics; *Scientific American*, 185, 4.
LEONTIEF, W. W., [1951B], *The Structure of the American Economy, 1919–1939*, (New York), 264 pp.
LEONTIEF, W. W. et al, [1953], *Studies in the Structure of the American Economy*, (New York), 561 pp.
LEONTIEF, W. W., [1965A], The Structure of the U.S. Economy; *Scientific American*, 212 (4), 25–35.
LEONTIEF, W. W., [1965B], The Structure of Development; In *Technology and Economic Development*, (Harmondsworth, Middlesex), 129–148.
LOMBARDINI, S., [1964], Les Analyses Economiques pour la Préparation d'un Plan Régionale; *Revue D'Economie Politique*, 74 (1), 45–64.
LONSDALE, C., [1965], Planning Britain's Regions; *Town and Country Planning*, 33 (2), 83–90.
LÖSCH, A., [1954], *The Economics of Location*, (New Haven), 520 pp.
LUTZ, V., [1960], Italy as a Study in Development; *Lloyds Bank Review*, 58, 31–45.
MABOGUNJE, A. L., [1965], Urbanization in Nigeria – A Constraint on Economic Development; *Economic Development and Cultural Change*, 8 (4), Pt. 1, 413–438.
MAGEE, B., [1965], *Towards 2000*, (London), 156 pp.
MAHALANOBIS, P. C., [1963], *The Approach of Operational Research to Planning in India*, (Bombay), 168 pp.
MAKI, W. R. and TU, YIEN-I, [1962], Regional Growth Models for Rural Areas Development; *Papers and Proceedings of the Regional Science Association*, 9, 235–244.
MANNERS, G., [1962], Regional Protection: a factor in Economic Geography; *Economic Geography*, 38 (2), 122–129.
MANNERS, G., (Ed.), [1964], *South Wales in the Sixties*, (Oxford), 265 pp.
MARTIN, W., [1957], Ecological Change in Satellite Rural Areas; *American Sociological Review*, 22 (2), 173–183.
MCGOVERN, P. D., [1965], Industrial Dispersal; *Planning*, 31, 485, 39 pp.
MCLAUGHLIN, G. E. and ROBOCK, S., [1949], Why Industry Moves South; *National Planning Association, Committee of the South*, Report No. 3, 148 pp.
MCLOUGHLIN, P. F. M., [1966], Development Policy-Making and the Geographer's Regions: Comments by an Economist; *Land Economics*, 42 (1), 75–84.
MCNEE, R. B., [1959], The Changing Relationships of Economics and Economic Geography; *Economic Geography*, 35 (3), 189–198.
MEIER, G. M., [1953], Economic Development and the Transfer Mechanism; *Canadian Journal of Economics and Political Science*, 19 (1), 1–19.
MEIER, G. M., [1964], *Leading issues in development economics*, (New York), 572 pp.
MEIER, G. M. and BALDWIN, R. E., [1957], *Economic Development. Theory, History, Policy*, (New York), 588 pp.
MELAMID, A., [1955], Some Applications of Thunen's Model in Regional Analysis of Economic Growth; *Papers and Proceedings of the Regional Science Association*, 1, L1–L5.
MEYER, J. R., [1963], Regional Economics: A Survey; *American Economic Review*, 53 (1), 19–54.

MEYERS, F., [1965], *Area Redevelopment Policies in Britain and the Countries of the Common Market*, (Los Angeles).

MEYNAUD, J. (Ed.), [1963], *Social Change and Economic Development*, (Paris), 210 pp.

MILHAU, T., [1956], La Théorie de la Croissance et L'Expansion Régionale; *Economie Appliquée*, 9 (3), 349–366.

MILLS, G., [1964], Economics and Computers; *The Technologist*, 1 (4), 15–22.

MOORE, F. T. and PETERSEN, J. W., [1955], Regional Analysis: An Interindustry Model of Utah; *Review of Economics and Statistics*, 37 (4), 368–383.

MOORE, F. T., [1955], Regional Economic Reaction Paths; *American Economic Review*, 45 (2), 133–148.

MOSES, L. N., [1955], Location Theory, Input-Output, and Economic Development: An Appraisal; *Review of Economics and Statistics*, 37 (3), 308–312.

MOUNTJOY, A. B., [1963], *Industrialization and Under-Developed Countries*, (London), 223 pp.

MUNGER, E. S., [1965], *Bechuanaland*, (London), 114 pp.

MYINT, H., [1964], *The Economics of Developing Countries*, (London), 192 pp.

MYRDAL, G. M., [1956], *Development and Underdevelopment*, (Cairo), 88 pp.

MYRDAL, G. M., [1957A], *Rich Lands and Poor: the road to world prosperity*, (New York), 168 pp.

MYRDAL, G. M., [1957B], *Economic Theory and Under-Developed Regions*, (London), 168 pp.

NATIONAL ECONOMIC DEVELOPMENT COUNCIL, [1963], *Conditions Favourable to Faster Growth*, (London), 54 pp.

NEEDLEMAN, L. and SCOTT, B., [1964], Regional Problems and Location of Industry Policy in Britain; *Urban Studies*, 1 (2), 153–173.

NEUMARK, D. S., [1964], Foreign Trade and Economic Development in Africa; *Stanford Food Research Institute, Miscellaneous Publications* 15, 222 pp.

NICHOLLS, W. H., [1961], Industrialization, Factor Markets, and Agricultural Development; *Journal of Political Economy*, 69 (4), 319–340.

NICHOLSON, I., [1965], *The X in Mexico: Growth within Tradition*, (London), 319 pp.

NORTH, D. C., [1955], Location Theory and Regional Economic Growth; *Journal of Political Economy*, 63 (3), 243–258.

NORTH, D. C., [1956], A Reply; *Journal of Political Economy*, 64 (2), 165–168.

NORTH, D. C., [1958], A Note on Professor Rostow's 'Take-Off' Into Self-Sustained Economic Growth; *Manchester School of Economic and Social Studies*, 26 (1), 68–75.

NORTH, D. C., [1959], Agriculture in Regional Economic Growth; *Journal of Farm Economics*, 41 (5), 943–951.

NORTH, D. C., [1961], *The Economic Growth of the United States, 1790–1860*, (Englewood Cliffs), 304 pp.

NURKSE, R., [1961], *Patterns of Trade and Development*, (Oxford), 62 pp.

O'CONNOR, A. M., [1963], Regional Contrasts in Economic Development in Uganda; *East African Geographical Review*, 1, 33–43.

OHLIN, G., [1961], Reflections on the Rostow Doctrine; *Economic Development and Cultural Change*, 9 (4), Part I, 648–655.

OKUN, B. and RICHARDSON, R. W., [1961], Regional Income Inequality and Internal Population Migration; *Economic Development and Cultural Change*, 9 (2), 128–143.

OOI JIN-BEE, [1959], Rural Development in Tropical Areas, with Special Reference to Malaya; *Journal of Tropical Geography*, 12, 1–222.

ORCHARD, J. E., [1960], Industrialization in Japan, China Mainland, and India – Some World Implications; *Annals of the Association of American Geographers*, 50 (3), 193–215.

ORCUTT, G. H., [1960], Simulation of Economic Systems; *American Economic Review*, 50 (5), 893–907.

PAAUW, D. S., [1961], Some Frontiers of Empirical Research in Economic Development; *Economic Development and Cultural Change*, 9 (2), 180–199.

PAHL, R. E., [1965], Urbs in Rure; *London School of Economics and Political Science, Geographical Papers* 2, 83 pp.

PAISH, F. W., [1964], The Two Britains; *The Banker*, 114, 456, 88–98.

PARR, J. B., [1965], *The Nature and Function of Growth Poles in Economic Development*, (Seattle), mimeo., 13 pp.

PARR, J. B., [1966], Outmigration and the Depressed Area Problem; *Land Economics*, 42 (2), 149–159.

PEN, J., [1965], *Modern Economics*, (Harmondsworth), 266 pp.

PERLOFF, H. S., et al., [1960], *Regions, Resources and Economic Growth*, (Baltimore) 716 pp.

PERLOFF, H. S., and WINGO, L., [1961], Natural Resource Endowment and Regional Economic Growth; In Spengler, J. J. (Ed.), *Natural Resources and Economic Growth*, (Washington, D.C.), 191–212.

PERROUX, F., [1955], Note sur la notion de 'pôle de croissance'; *Economie Appliquée*, 8 (1–2), 307–320.

PERROUX, F., [1961], *L'économie du XXème siècle*, (Paris), 598 pp.

PFISTER, R. L., [1963], External Trade and Regional Growth: A Case Study of the Pacific Northwest; *Economic Development and Cultural Change*, 11 (2), Part I, 134–151.

POLITICAL AND ECONOMIC PLANNING, [1962], *Regional Development in the European Economic Community*, (London), 95 pp.

POLLOCK, N. C., [1960], Industrial Development in East Africa; *Economic Geography*, 36 (4), 344–354.

PRATTEN, C. and DEAN, R. M., [1965], The Economies of Large Scale Production in British Industry; *University of Cambridge, Department of Applied Economics, Occasional Paper*, 3, 105 pp.

PREBISCH, R., [1950], *The Economic Development of Latin America and its Principal Problems*, (Lake Success), 59 pp.

PREBISCH, R., [1959], Commercial Policy in the Underdeveloped Countries; *American Economic Review*, 49 (2), 251–273.

PRED, A., [1965], Industrialization, Initial Advantage, and American Metropolitan Growth; *Geographical Review*, 55 (2), 158–185.

RANDALL, L., [1962], Labour Migration and Mexican Economic Development; *Social and Economic Studies*, 11 (1), 73–81.

RANIS, G. and FEI, J. C. H., [1961], A Theory of Economic Development; *American Economic Review*, 51 (4), 533–565.

RAO, V. K. R. V., [1964], *Essays in Economic Development*, (London), 333 pp.

ROBINSON, E. A. G. (Ed.), [1960], *Economic Consequences of the Size of Nations*, (London), 447 pp.

ROBOCK, S., [1956], Regional Aspects of Economic Development, with special reference to recent experience in Northeast Brazil; *Papers and Proceedings of the Regional Science Association*, 2, 51–69.

ROBOCK, S. H., [1963], *Brazil's Developing Northeast: A Study of Regional Planning and Foreign Aid*, (Washington, D.C.), 213 pp.

RODWIN, L., [1955], Planned Decentralisation and Regional Development with Special Reference to the British New Towns; *Papers & Proceedings of the Regional Science Association*, 1, A1–A8.

ROMUS, P., [1958], *Expansion économique régionale et Communauté Européene*, (Leyden), 376 pp.

ROSTOW, W. W., [1955], *An American Policy in Asia*, (New York), 59 pp.

ROSTOW, W. W., [1956], The Take-Off into Self-Sustained Growth; *Economic Journal*, 66, 25–48.

ROSTOW, W. W., [1959], The Stages of Economic Growth; *Economic History Review*, Second Series, 12 (1), 1–16.

ROSTOW, W. W., [1960], *The Stages of Economic Growth: A Non-Communist Manifesto*, (Cambridge), 179 pp.

ROSTOW, W. W., (Ed.), [1963], *The Economics of Take-Off into Sustained Growth*, (London), 482 pp.

ROSTOW, W. W., [1964], *View from the Seventh Floor*, (New York), 178 pp.

ROYAL COMMISSION ON THE DISTRIBUTION OF THE INDUSTRIAL POPULATION, [1940], *Report*, (London), Cmd 6153, 320 pp.

RUTTAN, V. W., [1955], The Impact of Urban-Industrial Development on Agriculture in the Tennessee Valley and the Southeast; *Journal of Farm Economics*, 37 (1), 38–56.

RUTTAN, V. W., [1959], Discussion: the Location of Economic Activity; *Journal of Farm Economics*, 41 (5), 952–954.

SANDEE, J., [1959], *A Long-Term Planning Model for India*, (New York), 39 pp.

SCHUH, G. E. and LEEDS, J. R., [1963], A Regional Analysis of the Demand for Hired Agricultural Labor; *Papers & Proceedings of the Regional Science Association*, 11, 295–308.

SCHULTZ, T. W., [1950], Reflections on Poverty within Agriculture; *Journal of Political Economy*, 58 (1), 1–15.

SCHULTZ, T. W., [1953], *The Economic Organisation of Agriculture*, (New York), 374 pp.

SCHURR, S. H., [1965], Energy; In *Technology and Economic Development*, (Harmondsworth, Middlesex), 91–106.

SCOTTISH DEVELOPMENT DEPARTMENT, [1963], *Central Scotland. A Programme for Development and Growth*, (Edinburgh), 47 pp.

SICKLE, J. V. VAN, [1951], The Southeast: A Case Study in Delayed Industrialisation; *American Economic Review*, 41 (2), 384–393.

SICKLE, J. V. VAN, [1954], Regional Economic Adjustments: the Role of Geographical Wage Differentials; *American Economic Review*, 44 (2), 381–392.

SINGER, M., [1961], Cumulative Causation and Growth Economics; *Kyklos*, 14, 533–545.

SISLER, D. G., [1959], Regional Differences in the Impact of Urban-Industrial Development on Farm and Nonfarm Income; *Journal of Farm Economics*, 41 (5), 1100–1112.

SMITH, P. E., [1961], Markov Chains, Exchange Matrices, and Regional Development: *Journal of Regional Science*, 3 (1), 27–36.

SMITH, W., [1953], *An Economic Geography of Great Britain*, (London), 756 pp.

SMOLENSKY, E., [1961], Industrialisation and Income Inequality – Recent United States Experience; *Papers & Proceedings of the Regional Science Association*, 7, 67–88.

SNYDER, D. E., [1964], Alternative Perspectives on Brasilia; *Economic Geography*, 40 (1), 34–45.

SPENCER, J. E., [1960], The Cultural Factor in 'Underdevelopment': The Case of Malaya; Ch. 3 in Ginsburg, N. (Ed.), Essays on Geography and Economic Development, *University of Chicago, Department of Geography Research Paper*, 62, 173 pp.

SPENGLER, J. J., (Ed.), [1961], *Natural Resources and Economic Growth*, (Washington, D.C.), 306 pp.

STAMP, L. D., [1953], *Our Undeveloped World*, (London), 187 pp.

STAMP, L. D., [1963], *Our Developing World*, (London), 195 pp.

STEEL, R. W., [1964], Geographers and the Tropics; In Steel, R. W. and Protheroe, R. M. (Eds.), *Geographers and the Tropics: Liverpool Essays*, (London), 1–29.

STEVENS, B., [1958], Interregional Linear Programming; *Journal of Regional Science*, 1 (1), 60–98.

STOLPER, W., [1954–55], Spatial Order and the Economic Growth of Cities; *Economic Development and Cultural Change*, 3, 137–146.

STONE, R., [1956], Input-Output and the Social Accounts; Ch. 6 in Barna, T. (ed.), *The Structural Interdependence of the Economy*, (New York), 429 pp.

STONE, R., [1961], An Econometric Model of Growth: The British Economy in Ten Years Time; *Discovery*, 22 (5), 216–219.

STONE, R., [1964A], Computer Models of the Economy; *New Scientist*, 21, 381, 604–605.

STONE, R., [1964B], Mathematics in the Social Sciences; *Scientific American*, 211 (3), 168–182.

STONE, R., [1965], Social Accounting Matrix Models – a Framework for Economic Decisions; In Berners-Lee, C. M. (Ed.), *Models for Decision*, (London), 136–149.

STURMTHAL, A., [1955], Economic Development, Income Distribution, and Capital Formation in Mexico; *Journal of Political Economy*, 63 (3), 183–201.

SUPPLE, B. E., (Ed.), [1963], *The Experience of Economic Growth. Case Studies in Economic History*, (New York), 458 pp.

TANGRI, S., [1962], Urbanization, Political Stability, and Economic Growth; In Turner, R. (Ed.), *India's Urban Future*, (Berkeley), 192–212.

TATTERSALL, J. N., [1962], Exports and Economic Growth: The Pacific Northwest

1880 to 1960; *Papers & Proceedings of the Regional Science Association,* 9, 215-234.

THOMAS, M. D., [1963], Regional Economic Growth and Industrial Development; *Papers of the Regional Science Association, European Congress 1962,* 10, 61-75.

THOMPSON, W. R. and MATTILA, J. M., [1959], *An Econometric Model of Postwar State Industrial Development,* (Detroit), 116 pp.

TIEBOUT, C. M., [1956], Exports and Regional Economic Growth; *Journal of Political Economy,* 64 (2), 160-164 and 169.

TINBERGEN, J., [1959], Comparative Studies of Economic Growth; In National Bureau of Economic Research, *The Comparative Study of Economic Growth and Structure,* (New York), 193-200.

TINBERGEN, J., [1960], Regional Planning: Some Principles; *Netherlands Economic Institute, Division of Balanced International Growth, Publication 21/60,* 13 pp.

TINBERGEN, J., [1964], *Regional Planning,* (Rotterdam), 11 pp.

TINBERGEN, J. and BOS, H. C., [1962], *Mathematical Models of Economic Growth,* (New York), 131 pp.

TOSI, J. A. and VOERTMAN, R. F., [1964], Some Environmental Factors in the Economic Development of the Tropics; *Economic Geography,* 40 (3), 189-205.

ULLMAN, E. L., [1958], Regional Development and the Geography of Concentration; *Papers & Proceedings of the Regional Science Association,* 4, 179-198.

ULLMAN, E. L., [1960], Geographic Theory and Underdeveloped Areas; Ch. 2 in Ginsburg, N. (Ed.), *Essays on Geography and Economic Development; University of Chicago, Department of Geography Research Paper,* 62, 173 pp.

UNITED NATIONS, Department of Economic and Social Affairs, [1961], Use of Models in Programming; *Industrialization and Productivity,* 4, 7-17.

USHER, D., [1965], 'Equalizing Differences' in Income and the Interpretation of National Income Statistics; *Economica, New Series,* 32, 127, 253-268.

VALAVANIS-VAIL, S., [1955], An Econometric Model of Growth: U.S., 1869-1953; *American Economic Review,* 45 (2), 208-221.

VERBURG, M. C., [1964A], The Gent-Terneuzen Developmental Axis in the Perspective of the European Economic Community; *Tijdschrift voor Economische en Sociale Geografie,* 55 (6/7), 143-150.

VERBURG, M. C., [1964B], Location Analysis of the Common Frontier Zones in the European Economic Community; *Papers of the Regional Science Association, European Congress,* 12, 61-78.

VINING, R., [1946], The Region as a Concept in Business-Cycle Analysis; *Econometrica,* 14 (3), 201-218.

VINSKI, I., [1962], Regional Distribution of National Wealth in Yugoslavia; *Papers of the Regional Science Association, European Congress,* 3, 127-168.

WILHELM, W., [1950], Soviet Central Asia: Development of a Backward Area; *Foreign Policy Reports,* 217-228.

WILLIAMSON, J. G., [1965], Regional Inequality and the Process of National Development: A Description of the Patterns; *Economic Development and Cultural Change,* 13 (4), Part II, 84 pp.

WILSON, T., [1964], Policies for Regional Development; *University of Glasgow Social & Economic Studies, Occasional Papers,* 3, 93 pp.

WONNACOTT, R. J., [1964], Wage Levels and Employment Structure in United

States Regions: A Free Trade Precedent; *Journal of Political Economy*, 72 (4), 414–419.

WOOD, W. D. and THOMAN, R. S., (Eds.), [1965], *Areas of Economic Stress in Canada*, (Kingston, Ontario), 221 pp.

WRIGHT, M., [1965], Regional Development: Problems and Lines of Advance in Europe; *Town Planning Review*, 36 (3), 147–164.

WRIGLEY, E. A., [1965], Changes in the Philosophy of Geography; Ch. 1 in Chorley, R. J. and Haggett, P., (Eds.), *Frontiers in Geographical Teaching*, (London), 378 pp.

YATES, P. L., [1961], *El Desarollo Regional de Mexico*, (Mexico City), 405 pp.

CHAPTER NINE

Models of Urban Geography and Settlement Location

B. J. GARNER

INTRODUCTION

Settlement studies form a traditional part of Human geography. They have held a dominant place in early statements, like that of Brunhes (1925), and in contemporary reviews like that by Jones (1964). Forming such distinctive features in the landscape, they were viewed as a fundamental expression of 'Man-Land' relationships. Perhaps not surprisingly, studies have traditionally emphasized strong links between the physical environment and various aspects of human occupance of regions; patterns of settlement distribution and morphology were all too often 'accounted' for by physical features. With the realization that urban areas themselves are regions full of interest for study it was merely a matter of applying the traditional 'Man-Land' concepts to the untapped sources for geographical investigation. The 'Townscape' became the urban equivalent of landscape and attention was drawn immediately to microscopic differences in the 'feel' or character of the various parts of urban areas. Emphasis was placed on form and patterns were essentially viewed as a reflection of physical controls supplemented heavily by historical influences. Although there is no question of the usefulness of these kinds of study, new frameworks are needed if a deeper understanding of the spatial organization of this particular aspect of human activity is to be obtained.

Settlements are considered as comprising a complex set of 'Man-Man' relationships here and this concept underlies the models discussed in this chapter. The importance of physical agents is overshadowed by the emphasis placed on various economic and social factors. But the impressive literature dealing with the many aspects of settlements is too large to be included in its entirety here; we must be selective. Emphasis is given to static rather than dynamic or predictive models and little explicit attention is paid to intra-urban transport and movement models. The chapter is divided into two parts: (1) a review of the major models and empirical studies dealing with the location and arrangement of settlements, and (2) a discussion of some of the models pertaining to the internal structure of urban areas.

Some underlying regularities in models

Virtually all models of settlement location and urban structure have one thing in common; they assume a measurable degree of order in spatial behaviour. This seems to be founded on the following six premises which form the basis of, or are implied in, most models.

1. *The spatial distribution of human activity reflects an ordered adjustment to the factor of distance.* Distance is basic to geography. Watson (1955) even goes as far as to state that geography itself is a 'discipline in distance'. This is not hard to see, for if all things were concentrated at a given place at a given time, there would be no patterns, no spatial variation or areal differentiation, in short there would be no geography. However, regularities with distance may not be immediately apparent. Writing on *Locational Analysis in Human Geography*, Haggett (1965, p. 2) quotes Sigwart's words, '. . . that there is more order in the world than appears at first sight is not discovered till order is looked for'.

The search for order in spatial behaviour must be accompanied by greater flexibility in thinking about distance. Different distance measures can be justified in model building by the fact that different things are more or less relevant in different types of studies. For example, in movement models, travel time (Voorhees, 1955), transport costs (Harris, 1954) or road distances weighted according to different kinds of road surface (Garrison, 1956) have proved more important than linear distance. Non-linear distance measures are equally important in locational models; Olsson and Persson (1964) have used density-distance, Getis (1963) income-distance and Garner (1966) land value-distance to reveal order in the distribution of urban functions.

2. *Locational decisions are taken, in general, so as to minimize the frictional effects of distance.* This concept, generally known as the 'law of minimum effort' (Lösch, 1954, p. 184) or the 'principle of least effort' (Zipf, 1949), suggests that events reach their goal by the shortest route. In settlement models in particular, movement-minimization is fundamental to an understanding of the geometry of settlement patterns and intra-urban location.Its importance is reflected in the place accorded the circle as a theoretical trade area and city shape (Haggett, 1965, p. 48).

3. *All locations are endowed with a degree of accessibility but some locations are more accessible than others.* Accessibility is difficult to define explicitly, but the

term generally implies 'ease of getting to a place' (Forbes, 1964). As such it is a variable quality of location. In a technical sense, accessibility is a relative quality accruing to a piece of land by virtue of its relationship to a system of transport (Wingo, 1961, p. 26). In an operational sense, it is the variable quality of centrality or nearness to other functions and locations. Clearly the notion of accessibility is closely related to the concept of movement-minimization, especially when this is measured by the costs involved in overcoming distance. In this context, it is also generally accepted as the basis of the rent paid for and the value attached to sites in urban land use models.

4. *There is a tendency for human activities to agglomerate to take advantage of scale economies.* Scale economies mean the savings in costs of operation made possible by concentrating activities at common locations. In the organization of an industrial firm it is exemplified in mass production techniques. Proximal concentration of a number of firms also make savings possible. Agglomerations can thus be viewed as nodes in the economic landscape arising from centripetal forces in spatial organization. The concentration of activities to form settlements themselves can be viewed as a reflection of scale economies but more important perhaps are the various agglomerations within urban areas such as shopping centres and industrial districts. Residential zones can also be viewed as agglomerations for scale economies in household costs of utilities and public services.

5. *The organization of human activity is essentially hierarchical in character.* This is true of both spatial and non-spatial aspects of human activity. For example, it is true of political organization although it may not be explicitly expressed spatially. However, the latter is a fundamental aspect of the spatial structure of settlement patterns and appears to result from interrelationships between agglomeration tendencies and accessibility. More accessible locations appear to be the sites of larger agglomerations. One of the implications from this is that there exists in an area a hierarchy of locations in terms of accessibility.

6. *Human occupance is focal in character.* This notion underlies the concept of the nodal or functional region and is basic to movement models and the spacing of certain activities in areas. The nodes about which human activity is organized are agglomerations of varying size. Since these are hierarchically arranged it follows that there is a hierarchy of different sized focal regions. Philbrick (1957) argues that the areal structure of the occupance of the earth's surface is composed of a number of hierarchically 'nested' orders of spatial functional organization. In this way, movement-minimization, accessibility, agglomerations and hierarchies are linked together to form a system of human organization in space.

SETTLEMENT PATTERNS

The existence of varying sized population clusters in the landscape is an inevitable feature in the spatial organization of human activity. Settlements exist because certain activities can be carried on most efficiently if they are clustered together rather than dispersed. No matter what the particular activities are, they can be generally viewed as services which are provided not just for settlements themselves, but for people living in surrounding tributary areas. Since settlements are spatially separated one from another, linkages between them are essential, and one framework for study is to view them as nodes or focal points in a transport network. We are concerned in this section with models of the arrangement of these varying sized nodes. For convenience our analysis will be divided into two components: (1) a 'horizontal' component in which the spatial parameter is explicit. Focus here is on maps and concerns regularities in the size and spacing of settlements; (2) a 'vertical' or organizational component, in which the spatial parameter need *not* be explicit. Focus here is on regularities which appear on graphs. However, it must be realized that the two components are mutually interrelated in reality. For example, the size and spacing of settlements may affect, and give rise to, regularity in vertical organization and *vice versa* – our division is consequently merely one of expediency.

Simple horizontal arrangements

A map showing the pattern of settlement in a region can be broken down into three basic parts (Harris and Ullman, 1945); (1) a linear pattern consisting of transport centres performing break-of-bulk and allied services and for which location is related to the disposition of transport routes; (2) a cluster pattern, consisting of places performing specialized services such as manufacturing, mining or recreation, and for which location is related to the localization of resources, and (3) a uniform pattern, consisting of places whose prime function is the provision of a wide range of tertiary goods and services and for which location is related to a dispersed population.

So far no model has satisfactorily taken into account all three aspects of urban causation and location, although steps have been taken in this direction by Lösch (1954), Isard (1956) and Bunge (1962, p. 59). Much recent geographical research has used models developed to explain the distribution and arrangement of the third category of centres, commonly referred to as central places, as a framework for study, and it is with these that we will concern ourselves here.

A. *The regular lattice model.* This is the basis of the central place model of

Christaller (1933), which he viewed as a 'general deductive theory' to explain the 'size, number and distribution of towns' in the belief that 'there is some ordering principle governing the distribution' (Berry and Pred, 1961, p. 15). Although this claim may be too great, there is no doubt that the model has had a profound impact on geographical research in the last decade and Bunge (1962) at least views it as crucial for the existence of a theoretical geography. In view of the importance of Christaller's work, it is surprising that there was not a satisfactory translation from the original German until that by Baskin in 1966. Perhaps even more surprising is that there is no complete review of the ideas contained in the model although discussion of its spatial aspects can be found in Berry and Pred (1961, pp. 3–18), Ullman (1941), Berry and Garrison (1958A, B and C) and Bunge (1962), and a critique of its economic assumptions is given by Baskin (1966).

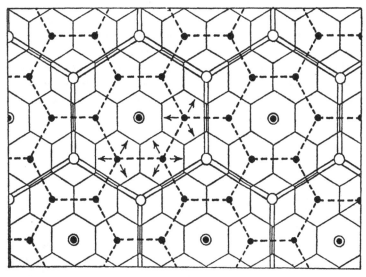

9.1 K=3 settlement pattern according to Christaller's marketing principle.

The basic elements of the regular lattice model, shown in Figure 9.1, are developed under assumptions of an isotropic surface – that is under conditions of a uniform distribution of population and purchasing power, uniform terrain and resource localization, and equal transport facility in all directions. In this ideal situation, the essential features of the 'Horizontal' arrangement of settlements are: (H-1) they are regularly spaced to form a triangular lattice, and (H-2) they are centrally located within hexagonal shaped trade areas. 'Vertical' organization hinges on the assumption that a hierarchy of discrete groups or orders of settlement exists in which (V-1) higher order places supply all the goods of lower order places plus a number of higher order goods and services that differentiates them from, and at the same time sets them above, central places of lower order, and (*V*-2) higher order places offer a greater range of goods and services, have more establishments, larger populations, trade areas and trade area populations and do greater volumes of business than lower order settlements. This 'vertical' organization has 'horizontal' expression in the following ways: (*C*-1) higher order central places are more widely spaced than lower order places, and (*C*-2) lower order central places, to be provided with higher order goods and services, are contained or 'nest' within the trade areas of higher order places according to a definite rule.

As shown in Figure 9.5, a hierarchy of settlements can be organized in various ways, each with its own geometrical arrangement of central places and trade area boundaries. In Christaller's basic model, organized on what he calls the marketing principle, the hierarchy and nesting pattern results in the maximum number of central places – a necessary condition if the supply of goods from central places is to be as near as possible to the consumer in accordance with the notion of movement-minimization. This particular system, known as a *K*-3 network, is the one shown in Figure 9.1. Three orders in the hierarchy are shown in the following way; (1) the lowest order places (e.g. hamlets) by filled circles, (2) intermediate order places (e.g. villages) by open circles, and (3) high order places (e.g. towns) by double circles. Trade area boundaries of the three orders of settlements are indicated by solid lines, dashed lines and double lines respectively.

The *K*-value, here three, refers to the number of settlements at a given level in the hierarchy served by a central place at the next highest order in the system. For example, in Figure 9.1, each village serves the equivalent of three hamlets. This number is made up of the hamlet part of the functional structure of the village itself (from *V*-1 above), plus a one-third share of the six border hamlets since each of them is shared between three villages as indicated by the arrows. Similarly, towns will provide town-level goods to three villages and it follows from the geometry of the hexagonal trade areas, they will serve nine hamlets. This regular progression, which can be extended upward from the town level, exists because Christaller assumed that once the *K*-value was adopted in any region, it would be fixed. Consequently, it would apply equally to the relationship between hamlets and villages, villages and towns, and so on up through the central place hierarchy. Because of this the total number of settlements in any area should follow a regular progression. In the case when the *K*-value equals three, this would be one, three, nine, twenty-seven, eighty-one . . . , starting with the highest order place in the region.

Although Christaller's *K*-3 model has received most attention in empirical studies, he did postulate two other forms of hierarchical arrangements to take account of deviations from the marketing principle. Firstly, a *K*-4 network (see Fig. 9.5) organized according to the traffic principle was proposed to account for situations in which costs of transport were significant. This arrangement enables '. . . as many important places as possible to lie on one traffic route between larger towns, the route being established as cheaply as possible' (Berry and Pred, 1961, p. 16), and gives rise to linear patterns and distortions of trade areas at right angles to traffic routes. 'Nesting' is according to the rule of fours because connections will be made with the equivalent of three of the six nearest dependent places. Secondly, a K-7 network was proposed to take account of the administrative principle or principle of 'separation', in which connections are made between a given order of central place and all six of the nearest immediately lower order places (see Fig. 9.5). Nesting is therefore according to the rule of sevens.

The literature dealing with various aspects of Christaller's models is both voluminous and contradictory. The reader is referred to the excellent annotated bibliography of central place studies by Berry and Pred (1961) and to Haggett (1965) for a summary of findings. Here we will look at the evidence for a regular lattice distribution of settlement in reality.

Hexagons: One of the commonest criticisms of the regular lattice and the hexagonal framework is that it is far too rigid and abstract. Although this is certainly true up to a point since the hexagon is a pure concept much as perfect competition is a pure concept to the economist, there are forceful theoretical arguments (see Haggett, 1965, p. 49, and Lösch, 1954, p. 105) for thinking in this framework if movement-minimization is at all relevant in spatial behaviour. It is not surprising to find, therefore, that very few studies have actually tested whether hexagonal arrangements do in fact exist in reality, and more are needed. In part this has perhaps been due to the subjective delimitation of trade areas (compare the methods used by Smailes (1947) with those by Green (1950) for example) and in part because until recent suggestions by Bunge (1962) and Boyce and Clark (1964) the concept of shape was difficult to quantify.

One study which provides evidence of hexagons in territorial organization is that by Haggett (1965, p. 50), who studied shape patterns of political division (*municipios*) in Brazil. Although the pattern was dominated by elongated shapes, the number of contacts between one territory and adjacent territories showed evidence of hexagonal arrangement. Nearly one out of three *municipios* bordered on exactly six neighbours while the mean number of contacting sides was 6·21 compared to six in the perfectly hexagonal system shown in Figure 9.1. Although this kind of investigation must be extended to less easily defined trade area boundaries before we can definitely say that the hexagon is a basic part of reality, the rather striking approximations of Haggett's findings to the model suggests that '. . . criticism of the hexagonal system as being over-theoretical may have been too hasty' (Haggett, 1965, p. 51).

Regular lattices: Until the recent introduction of a statistical definition of spatial uniformity based on nearest neighbour analysis (Clark and Evans, 1954; Dacey, 1963), it was difficult rigorously to measure dot-patterns. The more traditional 'eye-ball' methods are not really satisfactory. Nearest neighbour analysis enables patterns to be measured by an index, *Rn*, which ranges between zero for a perfectly clustered pattern and a maximum of 2·15 for a perfectly uniform pattern. A value of one is associated with random patterns. These three patterns are illustrated in Figure 9.2. Applications of this statistical measure to settlement patterns have shown them to tend toward randomness rather than the uniform distributions postulated in the model.

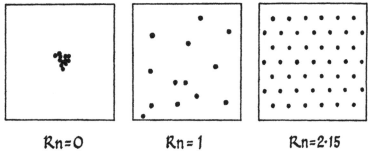

$Rn=0$ $Rn=1$ $Rn=2·15$

9.2 Clustered, random and perfectly uniform distributions.

Typical of studies using the statistical approach is that by King (1962) who analysed the settlement patterns in twenty sample areas in the United States (Fig. 9.3A). Comparison of the observed linear distances between a place and its nearest neighbour, regardless of size, with the spacing expected in a random distribution yielded a range in *Rn* values. As Figure 9.3B illustrates, these show a small range from 0·7 for Utah with a relatively clustered pattern to 1·38 for Missouri which tends towards a uniform spacing (Fig. 9.3C). The concentration of *Rn* values for the remaining areas between these limits indicates that the settlement pattern of the United States approximates a random more than the hypothesized uniform distribution. This finding concurs with that found for part of Wisconsin by Dacey (1962). Although size of centre was explicitly considered in this study, at no level in the hierarchy was a uniform distribution indicated.

The failure of the regular lattice model of settlement distribution is hardly surprising in view of the idealized conditions under which it was hypothesized. Haggett (1965) illustrates the distorting effects of agglomeration, resource localization and of time lags and it is not difficult to think of others. But it might be that the lack of evidence for the regular lattice in empirical work is due to shortcomings in study design and measurements used. Thus

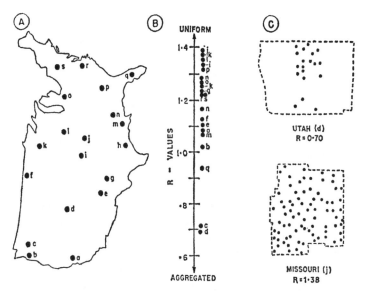

9.3 Sample study areas within the United States (A), scale of Rn-values (B), and clustered settlement patterns for Utah contrasted to uniform settlement pattern for Missouri (C) (*Source: Haggett, 1965, p. 91. After King, 1962, pp. 3–4*).

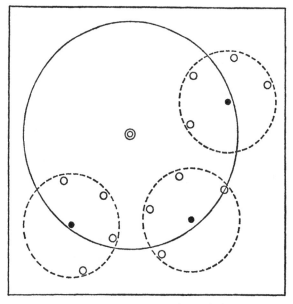

9.4 Regular cluster settlement pattern (*Source: Brush, 1953, p. 391*).

Dacey (1962) suggests that the notion of a lattice could be accepted for south-western Wisconsin but that the central place hierarchy on which he based his analysis had been incorrectly identified by Brush (1953) in the original study of the area. If this is so, then the definition of nearest neighbour is crucial in study design. Normally, measurements are made to the nearest neighbour of the *same* size. The problem is in deciding what constitutes a place of the same size, for apart from chance circumstances, two places are very unlikely to be exactly the same size, however measured. Rather, as pointed out by Thomas (1961), it is better to consider the nearest place of *approximately* the same size. To sharpen the definition of 'approximately', he introduces a statistical definition based on probability notions but the refinement has not so far been used to test for the regular lattice.

A more serious shortcoming concerns the measure of distance used. It can be argued that regularities are not found because the models have been tested under non-isotropic conditions for which they were never designed, and that to compensate for this disparity between 'model world' and real world, more functional measures of distance are wanted. Evidence that this might be worthwhile investigating is provided by Olsson and Persson (1964) who are able to explain a higher amount of variation in the spacing of settlements in Sweden by measuring distance as the number of people living between two places. This idea is consistent with a model of consumer movement proposed by Huff (1961) which postulates that shopping trips are restricted not only by physical distance but also by intervening population density. Although Huff's model was developed for intra-city movements there is no logical reason why similar arguments cannot apply to movements between cities and to their spacing.

Transformations of distance in this and other ways would be consistent with Christaller's original formulation which emphasized time-cost distance in the spacing of settlements. However, as Blome (1963) shows, there are shortcomings in the use of this distance transformation. A further possibility of eliminating the problem of non-isotropism of surface might be to abandon classical concepts of Euclidean geometry. Tobler (1963) has drawn attention to the possibilities for this and encouraging evidence is presented in a study by Getis (1963) which shows that when variations in income are smoothed out, uniform patterns are identified in the distribution of grocery shops in Tacoma city, USA.

B. *Regular cluster models.* An alternative hypothesis to the regular lattice model is that settlements form regular clusters. This is the basis of a model proposed by Kolb and Brunner (1946). In Figure 9.4, towns are shown by double circles, villages by filled circles and hamlets by open circles; solid lines indicate the trade area boundary of towns, and dashed lines the boundary of the influence of villages. The pattern is one in which (1) villages are located near the boundary of the influence of towns, (2) hamlets are clustered around villages at the edge of village trade areas, and (3) towns are centrally located.

The spacing of settlements is consequently one of clusters resulting from the influence of size on location. Although elements of the latter are included in the regular lattice model (e.g. C-1 above), the cluster model is less rigorous in that neither distances nor direction are specified and because of this it is perhaps more in accord with reality. Moreover, the model is in agreement with other models of spatial interaction including the 'Law of Retail Gravitation' (Reilly, 1931) and the 'Proportional Range of Influence' (Tuominen, 1949) which specify that the 'pull' exerted by a place varies directly with its size and decreases outwards with distance. Consequently smaller places are not likely to develop as close to large clusters as they are to one another.

Evidence from empirical studies of clustering is not conclusive and if anything is contradictory. Brush (1953) found hamlets crowding together in areas farthest from the larger towns in Wisconsin, USA, and Dacey (1962) confirmed this observation using nearest neighbour analysis. In other studies, for example that by Berry, Barnum and Tennant (1962) of central places in Iowa, USA, there is less evidence of clustering although it is shown that villages are functionally linked with cities rather than with towns, as envisaged in the model. Further evidence that smaller places grow up at the boundaries of trade areas between larger settlements is given for Sweden by Godlund (1956).

On the other hand, there is abundant evidence illustrating that the spacing of settlement is governed by the size of settlement. Brush and Bracey (1955) found from a comparative study of settlement patterns in south-western Wisconsin, USA and southern England that towns were spaced at an average distance of twenty-one miles apart and villages between eight and ten miles apart. Similar evidence of the wider spacing of larger settlements is given by Christaller (1933) for southern Germany and by Lösch (1954) for Iowa, USA. The implication from these studies is that regularities of this sort may well exist in many areas despite differences in population density, regional economy and social and economic history.

The general evidence of these earlier studies has been specified more exactly by Thomas (1961) for Iowa, USA, and Olsson and Persson (1964) for Sweden using regression analysis. These studies both show rather low, but statistically significant, correlations between size of centre and spacing. Only about a third of the variation in spacing could be 'explained' ($R^2 = 0.35$) in Iowa, and this seemed to be fairly stable over time (Thomas, 1962). In Sweden, a similar low 'explanation' ($R^2 = 0.31$) was found although this improved slightly ($R^2 = 0.46$) when the number of people living in theoretically delimited trade areas surrounding central places was used as a modification of distance, implying that the simple regularity is distorted by population densities not considered in the model. Although we can conclude from these studies that there is a significant relationship between size of city and spacing, it is obviously not as simple as postulated in the regular cluster model. The spacing of settlement must be considered in a wider context than size alone.

One of the few studies in which this has been attempted is that by King (1961) who used multiple regression analysis on a sample of 200 towns in the USA. His model specified that spacing was a complex function of size of centre, its occupational structure and the character of the region in which it is located. Table 9.1 shows that only two per cent of variation could be explained although, as to be expected, the spacing was more predictable for central places than for non-central places. The difficulties of generalizing over large

TABLE 9.1

Relationship between settlement spacing and other variables in the United States, 1950

Hypotheses:	Single hypothesis (Size of settlement)	Multiple hypothesis (Six-factors)
Coefficients of determination (R²):		
National results	0·02*	0·25*
Centre classification:		
Central places	0·09*	0·26*
Non-central places	0·01	0·42*
Regional agricultural classification		
Grazing and wheat zone	0·42*	0·67*
Specialized farming zone	0·01	0·20*
General farming zone	0·07	0·67*
Feed grain and livestock zone	0·22*	0·34*
Dairying zone	0·04*	0·36*

* Significant at the 95 per cent confidence level

Source: *King, 1961, pp. 227–31.*

areas such as the entire USA is also clearly revealed by the marked regional differences in spacing. In the Great Plains where physical conditions most closely approximate those assumed by Christaller, over forty per cent of variation in spacing could be explained compared to only four per cent in the more undulating country of the Dairy Region in the Middle West.

The low correlation between size of centre and spacing prompted King to investigate the effects of other factors on spacing. He argued that settlements of a given size were likely to be more widely spaced in areas of extensive farming, and where rural population densities, overall population densities, agricultural production and the proportion of workers in manufacturing were low. Regression analysis showed that all five variables were only slightly more valuable than town size in predicting spacing. Only overall population density explained as much as 10 per cent of the variation in spacing. Even when spacing was considered a function of all six variables taken together only one quarter of the variation in spacing could be accounted for, although the model gave a better fit when regional differences were considered.

Complex horizontal arrangements

Since Christaller's contribution, the hexagonal based model has been elaborated and extended by Lösch (1954) into a more complex model which approximates more nearly the patterns in the real world. The more flexible model is made possible by regarding the K-value as being free to vary. Thus

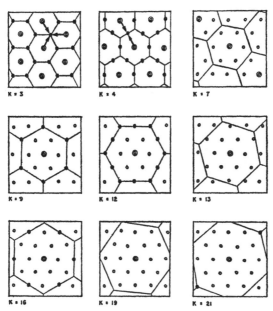

9.5 Nine smallest hexagonal trade areas in a Löschian landscape (*Source: Haggett, 1965, p. 119. After Lösch, 1954, p. 118*).

the fixed-K assumptions of the marketing, traffic and 'separation' principles of the Christaller model are considered to be special cases of a larger number of possible hexagonal systems and settlement distributions. The nine smallest feasible hexagonal arrangements associated with the various K-values are illustrated in Figure 9.5. These are obtained by varying the orientation and size of the hexagon. In the figure, high order places are shown by a double circle, the dependent lower order places by open circles if they lie completely within the trade area of the high order centre and by filled circles if they lie on the perimeter of it.

The nine arrangements shown in Figure 9.5 and further nets up to K-25 derived in a similar way are superimposed on one another in Figure 9.6 so that they all have the same central point. This point, Lösch argues, will be the site of a metropolis – the largest order of central place in his system. By rotating the nets about this point it is possible to derive a pattern which has six sectors with many, and six sectors with few, settlements. The sectoral

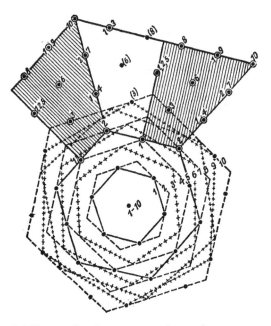

9.6 Ten smallest hexagons superimposed over a common central point. The shaded sectors contain many centres. Simple points represent original settlements; numbers beside encircled points indicate the size of trade areas (*Source: Lösch, 1954, p. 118*).

pattern, called an 'economic landscape', is shown in Figure 9.7. In this arrangement, Lösch claims that the greatest number of locations coincide, the aggregate distance between all settlements is minimized, and the maximum number of goods can be locally supplied.

The basic features of the 'horizontal' arrangement of settlement in the Lösch model are; (1) concentration of settlement into sectors separated by interstitial areas in which settlement is less dense, as shown in Figure 9.8A. Within the 'city rich' sectors, (2) settlement increases in size with distance from the central metropolis, as shown in Figure 9.8B, and (3) small settlements are located 'about half way' between two larger ones. This is shown in Figure 9.8C by the two larger classes of dots, representing places which are the centres of over four and over eight coincident hexagonal nets respectively.

Lösch further asserts, but does not demonstrate, that settlements comprise

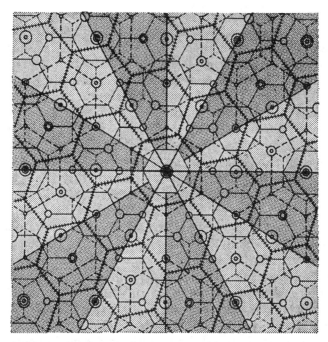

9.7 Simplified Löschian landscape and system of hexagonal trade areas (*Source: Haggett, 1965, p. 123. After Isard, 1956, p. 270*).

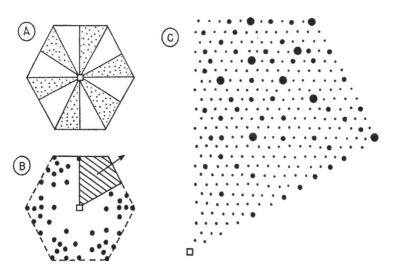

9.8 Alternating city-rich and city-poor sectors (A), distribution of large cities (B), and pattern of settlement within one sector (C) (*Source: Haggett, 1965, p. 123. After Lösch, 1954, p. 127*).

6

a hierarchy. Since this is based on a variable-K rather than the fixed-K of Christaller, 'vertical' organization is far less rigid. Its basic features are; (1) a continuous array of centres in which, (2) higher order places do not necessarily provide all the functions typical of lower order places, and (3) settlements performing the same *number* of functions do not necessarily provide the same *kinds* of functions. For example, because of the rotation of nets, a settlement serving seven smaller places may be either a K-7 settlement or merely the coincident centre for both a K-3 and a K-4 size market area.

Partial evidence in support of Lösch's 'city rich' and 'city poor' landscape is presented by Bogue (1949) from a study of population densities around the sixty-seven largest metropolitan centres in the USA. Bogue divided the hinterlands of each metropolitan area into three types of sector; (1) *route* sectors containing major highways leading from the central metropolis to other metropolitan centres, (2) *subdominant* sectors, which contained at least one city of more than 25,000 inhabitants, and (3) *local* sectors, which contained neither large cities nor inter-metropolitan routes. Figure 9.9E shows that urban densities are greatest in the subdominant sectors. Densities are lower than might be expected from the model in the route sectors (Fig. 9.9D), while local sectors were well below the level of the other two (Fig. 9.9F).

Bogue's study clearly revealed a logarithmic decline in population densities with distance away from the central metropolis. In Figure 9.9 this is shown by the broken line in all the graphs. At 25 miles from the city densities are over 200 per square mile whereas at 250 miles from it, densities have fallen to only about 4 persons per square mile. In addition, the findings suggest that densities are greater around larger cities than smaller ones, presumably a reflection of the differential forces of agglomeration, and that density patterns are markedly different from one region to another in the United States, (Fig. 9.9A, B and C).

Although partly confirming the Lösch model, Bogue's study points up one of its major deficiencies, namely the disparity between the postulated uniform size of hexagonal market areas and the implied decline in density of population with distance from the central metropolis. Under the latter conditions, market areas should be smaller close to the metropolis and should increase in size with distance from it. It follows from this that settlements should be more widely spaced the further they are away from the metropolis. However, this part of Lösch's work is not at all clear and more detailed information needs to be supplied, although Isard (1956, p. 272) has attempted a graphic modification, as shown in Figure 9.10, to take into account density differences. Considerable difficulty appears to have been encountered in trying to adhere to the hexagonal shape of market area and in view of his criticisms of the Lösch model (Isard, 1956, p. 271) it is not clear why this basic shape was retained.

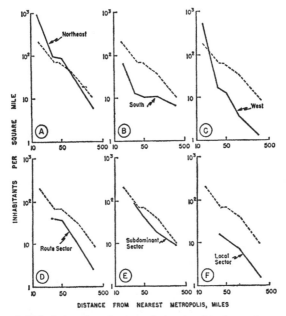

9.9 Variations in urban population densities by region
(A, B, C) and by sector (D, E, F) in the United States.
Both axes are transformed to logarithmic scales
(*Source: Haggett, 1965, p. 93. After Bogue, 1949, pp. 47
and 58*).

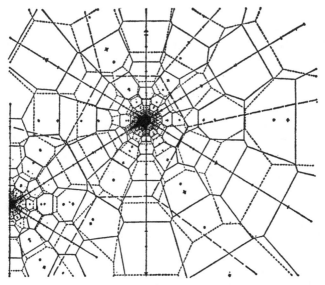

9.10 Löschian system of hexagonal trade areas modified by
agglomeration (*Source: Isard, 1956, p. 272*).

Periodic central places

In many parts of the world, especially the lesser developed countries, consumers are not normally supplied with goods and services from permanent central places. Rather, central functions are performed by mobile agents and their physical facilities moving from place to place during the short run of time. Such spatial behaviour of individual producers gives rise, in the aggregate, to the 'fair' or periodic market. Two or more merchants meeting at the same place at the same time are apt to sell more than if the time of their visits to a place is not co-ordinated. Thus, a regular schedule of markets emerges as the important feature of the central place system.

Even where permanent central places exist in the landscape, the number and kinds of functions they perform pulsates periodically when they become the site of markets. This is true, for example, of many small rural settlements in England today which still hold regular weekly markets. On such days, the normal hierarchy of central places is temporarily disturbed; low order places, enlarged by the addition of other functions comprising the market, assume higher order status. Concomitant with their increased importance, they serve larger trade areas than normal. In England this is reflected in the special bus services provided for the people living in surrounding rural areas on market days. In other areas of the world, it is reflected in the convergence of people from rural areas on the market centre. In both cases consumers, by submitting themselves to the discipline of time in trip-making, are able to free themselves from the discipline imposed by distance in order to purchase the goods they need.

An interesting model which accounts for the periodicity of central places and the behaviour of mobile merchants has been proposed by Stine (1962). The model is founded on the notions of the range of a good and the threshold. The maximum range of a good may be regarded diagrammatically as the radius generating a circle within which all consumers are willing to purchase at least some of a particular commodity. The threshold, or minimum range of a good, can be similarly viewed as the radius of a circle, but one containing the minimum demand necessary for the firm to earn normal profits, to stay in business and therefore become a viable entity in the economic landscape. Under normal conditions, these ranges are spatially related as shown in Figure 9.11A. The outer range, shown by the solid line, lies beyond the inner range. The establishment providing the particular good is permanent and has a fixed location. However, in areas characterized by a relatively primitive state of transport, consumers are unable to travel very far to buy goods because of the time needed to overcome the 'friction of distance'. Under these conditions, the position of the two ranges is likely to be reversed as shown in Figure 9.11B. When this situation pertains, there are two alternatives: (1) the firm cannot survive at a fixed location and so disappears from the land-

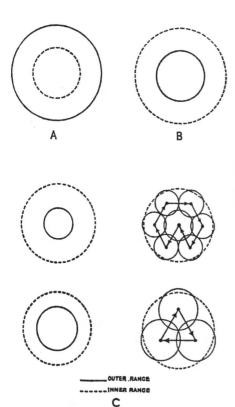

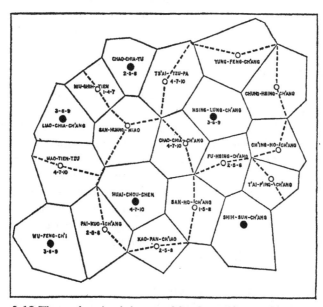

9.11 Normal (A) and inverted relationships (B) of the outer and inner ranges of a good. The number and length of paths for a mobile firm is a function of the radii of the two ranges of a good (C) (*Source: Stine, 1962, p. 22*).

————— OUTER RANGE
- - - - - INNER RANGE

C

9.12 The market circuit in part of Szechwan (*Source: Skinner, 1964, p. 22*).

scape, or (2) the firm must become mobile and move from place to place. When the latter course is decided upon, the number and length of the moves that the firm must make in order to stay in business would seem to be, as shown in Figure 9.11C, a function of the difference between the length of the radii for the maximum and minimum ranges for the particular good in question.

A good illustration of the periodic nature of central places has been provided recently by Skinner (1964) from a study of marketing in rural China. It appears that both the movement paths of travelling merchants and the spatial pattern of market locations support many of the basic elements of Christaller's central place model. For example, Figure 9.12 shows the location of six 'intermediate level' permanent central places (shaded circles) in the hierarchy of settlement in part of Szechwan. The places at which periodic markets are held according to a regular calendar are shown by open circles. The links between the larger settlements and the location of the periodic markets are shown by the dotted lines. Merchants travel from their base at an intermediate centre to six of the nearest surrounding smaller places. Since each of these is shared between two intermediate centres, a K-4 system seems to be typical of this area, but Skilling also shows that a K-3 system is typical of other parts of China.

These findings are interesting since they suggest that the basic notions of central place theory may have much wider application than was originally thought by Christaller, or has hitherto been imagined by recent researchers. The ideas would seem to provide a particularly good framework for the study of the weekly markets typical of English rural settlement.

Vertical arrangements

Hierarchies. It is a common observation that there are fewer larger places than smaller ones in a region and that the larger centres provide a greater number and variety of goods than the small places do. As we have seen this hierarchical organization is fundamental to the central place models of Christaller (1933) and Lösch (1954). More recently, Berry and Garrison (1958B) have demonstrated by using the notion of threshold how hierarchies arise in the landscape. Their model is more flexible since it does not assume a uniform distribution of population or a hexagonal system of trade areas. Thus it can be applied more widely and is especially important because it provides a logical explanation of a hierarchy of shopping centres, the equivalent of central places, within urban areas.

The relevant 'vertical' aspects of the model have been simplified and diagrammatically represented in Figure 9.13. The argument can be simply stated as follows. Figure 9.13A shows five goods with their respective threshold sizes; function A requires only 10 people to support it whereas function

E requires 160 people. Let us assume that these goods are to be supplied for an area with only 160 inhabitants. Once the population to be served and threshold sizes for each function are specified, the number of each function appearing in the landscape is determined. Figure 9.13B shows the numbers of each function in the example. Since the threshold of function *E* is equal to the total population to be served, only one of it can be provided. Conversely, the area can support 16 establishments of function *A* with its lower threshold

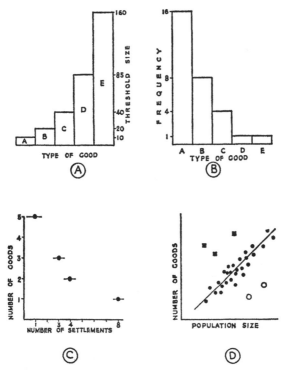

9.13 Threshold sizes for five functions (A), number of each type of function (B), size and number of settlements (C) and the simplified relationship of number of functions to size of settlement (D).

of 10 people. The numbers of each other good can be calculated in the same way. As a result some functions will be provided more frequently in the landscape than others. These are goods that must be provided conveniently (e.g. grocers). Other goods are supplied less frequently in the landscape, and are typical of goods for which the demand is more periodic (e.g. men's clothing shops). If a high order place supplies all lower threshold functions, then the number and size of settlements can be calculated. In this example, as Figure 9.13C shows, there will be a total of sixteen places; the largest provides all five functions, three supply goods *A*, *B* and *C*, four places supply the two lowest threshold functions, and eight places provide only function *A*.

A stepped distribution of different size settlements results. From (*V*-2) above, it was stated that higher order centres have larger populations. The regularity between functional size and population size, and two kinds of deviant cases are shown in Figure 9.13D. Some places may have fewer functions than expected from their population size (circles in Figure 9.13D) and thus occur below the line of the general relationship. Berry (1960) provides evidence that this is typical in hierarchies around large cities where effects of suburbanization give rise to 'dormitory' settlements lacking a complete range of functions. Alternatively, some places may have more functions than expected from their population size, and thus fall above the line of general relationship. These are shown in Figure 9.13D as squares. Thomas (1961) has identified deviants of this sort in a recreational area of northwest Iowa where central places, in catering for tourists, have a range of additional functions which are not characteristic of other Iowan cities.

Although hierarchies have been identified in many areas, for example by Brush (1953) in south-western Wisconsin, USA, by Bracy (1962) in southern England, by Berry, Barnum and Tennant (1962) in part of Iowa, and by Mayfield (1962) in India, there is considerable controversy as to whether in fact such hierarchies may be identified empirically or whether instead only a continuous functional relationship exists. Like all classificatory systems, it could be argued that the levels they postulated are merely ones of convenience for handling empirical data. Vining (1955, p. 169) summarizes this when saying, 'Like pool, pond and lake, the terms hamlet, village and town are convenient modes of expression, but they do not refer to structurally distinct entities'.

Perfect uniformity is rarely obtained in the real world. The relationship between number of functions and population size varies. Berry and Mayer (1962) argue that for large areas such variations are enough to provide the appearance of a continuous linear relationship even though the underlying spatial pattern may be that of a hierarchy, while Beckman (1958) has shown how, with the addition of random elements, the discrete steps of Christaller's hierarchy can be blurred into a continuous relationship. Moreover, it would appear that aggregate analysis inevitably emphasizes the importance of continuous arrangements, whereas elemental investigations usually identify a hierarchy.

Evidence of continuous relationships between the population and functional size of settlement in agreement with Lösch's model provided by Stafford (1963) for southern Illinois and Gunawardena (1964) for the southern part of Ceylon is shown in Figure 9.14. In both areas high positive correlations are found; ($R^2 = 0.79$) in the former area and ($R^2 = 0.83$) in the latter. A slightly lower correlation was found by Berry and Garrison (1958A) for Snohomish

County, Washington, USA ($R^2 = 0.55$). These findings confirm the hypothesis that larger places provide a greater range of functions than smaller places and, in addition, it appears from the curvilinear form of the relationship that as settlements become larger they add fewer functions for each additional increment in population size.

Evidence of breaks in the overall continuous relationship more in line with the Christaller hypothesis is provided by Berry and Garrison (1958A) for

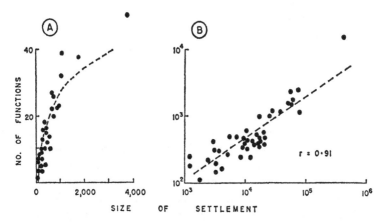

9.14 Relationships of functional size to settlement size in southern Illinois, USA (A), and southern Ceylon (B). The y-axis is transformed to a logarithmic scale in the second graph (*Sources: Haggett, 1965, p. 115. After Stafford, 1963, p. 170; Gunawardena, 1964*).

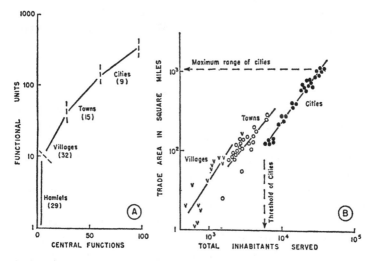

9.15 Four orders of central places in the settlement hierarchy, southwestern Iowa, USA (*Source: Haggett, 1965, p. 118. After Berry, Barnum and Tennant, 1962, pp. 79 and 80*).

Snohomish, County, Washington and Berry, Barnum and Tenant (1962) for south-western Iowa. In the latter study, factor analysis was used to group settlements on the basis of numbers and kinds of functions, and numbers of outlets (functional units). As Figure 9.15A shows, three distinct classes were recognized: (1) *cities* with more than 55 functions, (2) *towns* with from 28–50 functions, and (3) *villages* with between 10 and 25 functions. The lowest level of places, the *hamlets* were not included in the factor analysis. The horizontal expression of this vertical organization can be inferred from Figure 9.15B where each order of central place is shown on a logarithmic graph in relation to the size of trade area and total population served. In spite of overlap in the graph there is general agreement with the Christaller model in that higher order places serve larger tributary areas and populations than places at lower levels in the hierarchy.

Rank-size regularities. During the past fifty years interest has centred on another regularity in the size distribution of cities. This is the graphical relationship between the larger number of smaller places and fewer number of larger places in a region. When all cities in an area are ranked in decreasing order of population size, the size of a settlement of a given rank appears to be related to the size of the largest, or primate city, in the region. This regularity, generally known as the 'rank-size' rule, can be expressed by the formula,

$$Pr = Pl/r^q$$

where Pr is the population size of the city of rank r in the descending array of towns, Pl is the population of the largest city. This formula simply states that we should expect the population of the largest town to be four times the size of the fourth largest town in the region. Moreover, when the relationship is expressed in its logarithmic form

$$\log Pr = \log Pl - q \log r$$

it can be considered as lognormally distributed and a plot of rank on the x-axis against city size on the y-axis on double logarithmic graph paper should give a straight line with a slope $-q$ as shown by the upper line in Figure 9.16A.

Some of the alternative explanations of this empirical regularity are reviewed by Berry and Garrison (1958D). The most acceptable, though by no means easily grasped, of these seems to be that offered in terms of general systems theory by Simon (1955). Noticing that the form of the rank-size frequency distribution is identical to many probability distributions, among them the Yule and Lognormal, Simon argues that the regularity of city sizes is generated by some stochastic process. The close agreement between observed and expected frequency distributions in a recent study by Berry and Garrison (1958D) tend to confirm the Simon model as a plausible explanation of rank-size regularities.

This city relationship has been noted for many areas by as many writers. The formidable empirical evidence presented has been reviewed by Vining (1955) and Isard (1956, pp. 55–60); and Haggett (1965, p. 101), in summarizing the general evidence, concludes that although it may not be borne out exactly in reality, the rank-size regularity provides a useful framework within which generalizations about the population distribution of a region can be made.

The work of Stewart (1958) is perhaps typical of the application of the rank-size idea to large cities. From comparisons of the populations of the

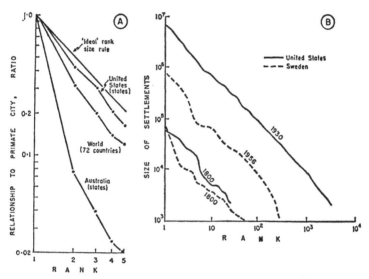

9.16 Median size of cities as a ratio of the largest city (A) and changes in the size-distribution of cities in the USA and Sweden (B) (*Source: Haggett, 1965, p. 102. After Stewart, 1958, pp. 228 and 231; Zipf, 1949*).

largest and second largest cities in seventy two different countries he concluded that the relationship was more varied than expected from the formula. For all countries sampled, the largest city was characteristically three and a quarter times, rather than twice, the size of the second largest city. Ratios varied from as high as 17·0 for Uruguay to 1·5 for Canada. The only regularity that Stewart could find was that larger countries had lower ratios.

Variation in rank-size relationships also appears to be characteristic of places lower down the urban spectrum. Figure 9.16A shows the reasonably close correspondence with the expected sequence for the five largest cities in the United States, and the strong divergence in Australia. Even when the rank-size relationship is extended over the full range of towns, straight line relationships are not always found. The curves shown in Figure 9.16B show two contrasted cases; (1) the linear form for the United States which generally conforms to the rank-size rule and (2) the irregular curve for Sweden. Comparison over time shows increasing linearity for the United States but increasing irregularity for Sweden.

Divergence from the expected straight line relationship seems to be due to a wide number of factors. Zipf (1949) suggests that the regularity is only characteristic of complete regions; that is, areas which are self-contained and not part of a larger region. Thus it could not be expected to hold for an English county but might for the entire British Isles. Stewart (1958) suggests that divergence from the rule is greater for homogeneous fairly well populated, mainly agricultural countries in which there are many smaller cities. Conversely, the rule holds well for industrialized countries principally because of the large size of some industrial centres, for areas with high rural population densities and for areas where population is well distributed spatially.

A recent study by Berry (1961) suggests, however, that it is not quite so

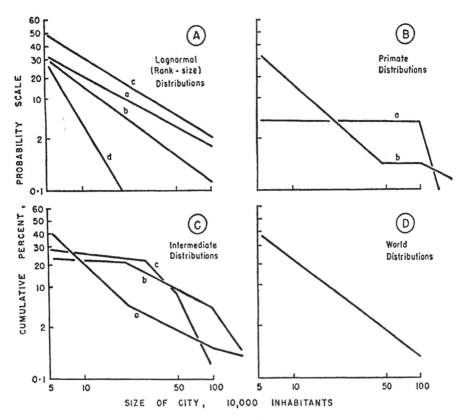

9.17 Alternative forms of city size distributions (*Source: Haggett, 1965, p. 104. After Berry, 1961, pp. 575–578*).

easy to relate the existence of rank-size relationships with general characteristics of a country. Berry plotted for each of thirty-eight countries the number of centres of over 20,000 inhabitants as a cumulative percentage on a normal probability scale on the y-axis and the size of centres on a logarithmic scale on the x-axis of a graph. Countries could be classified into three types according to their city size distributions. Firstly, 13 with *lognormal* (rank-size) distributions which appear as straight lines on this type of graph (Fig. 9.17A). This group included both highly developed countries like the United States (a) and underdeveloped countries like Korea (b); large countries like China (c) and small ones like El Salvador (d). Secondly, 15 countries with *primate* distributions in which a stratum of small towns and cities is dominated by one or more very large ones but in which there are very few centres of intermediate size. On the graph these appear as curved lines as shown in Figure 9.17B. Although all countries in this group are small, they show marked differences in types of curve. For example Thailand (a) lacks any signs of the lognormal curve, whilst Denmark (b) shows evidence of the lognormal distribution for its smaller cities. Thirdly, 9 countries showed *intermediate* distributions (Fig. 9.17C) including England and Wales (a), Australia (b) and Portugal (c). Figure 9.17D shows the tendency for the lognormal distribution to hold for all cities in the world.

As far as is possible to generalize from these graphs, lognormal (rank-size) distributions appear to be typical of larger countries which have long traditions of urbanization and which are politically and economically complex, whereas the exact converse holds for countries with primate distributions. They are generally smaller, have shorter traditions of urbanization and are economically and politically simple. More importantly perhaps, lognormal distributions are not only characteristic of technically advanced countries neither are primate distributions related in any significant way with underdeveloped countries as might intuitively be expected.

City classifications. When we refer to Sheffield as a 'steel' town or to Stoke-on-Trent as a 'pottery' town, we are recognizing a classification of towns by the economic functions they perform. Although all towns may be considered in part as central places for the provision of goods and services to local surrounding areas, for many this is subordinate to other specialized functions they perform for the wider regional and national markets. Groups of towns with similar functional specialization have been most frequently identified from the analysis of employment or occupation data. Specialization is said to exist when employment in a given industry category exceeds some 'normal' level. It is only when an abnormally large proportion of the labour force is employed in a particular activity that it becomes a distinguishing feature in differentiating that town from others.

Vital to the classification of towns is the problem of defining the 'normal' level of employment for a given industry group. In a number of studies, typified by the classification of American cities by Harris (1943), the break-point has been identified intuitively from observations of the employment-occupation structure of towns of well-defined types. The criteria used by Harris for identifying towns specializing in manufacturing, graphically presented in Figure 9.18A, are: (1) employment in manufacturing must equal at least 74 per cent of the total employment in manufacturing, retailing and wholesaling (y axis), and (2) manufacturing and mechanical industries must contain at least 45 per cent of the total gainfully employed labour force (x axis). Towns for which employment-occupation figures satisfy both of these criteria fall in the shaded upper right quadrant of the graph. Similar procedures using arbitrarily selected criteria were used to identify eight other types of specialized cities.

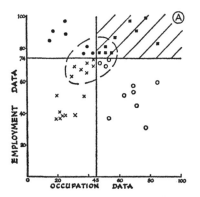 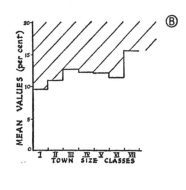

9.18 Methods of city classification used by Harris (A) and Pownall (B) (*Source: Harris, 1943, p. 87; Pownall, 1953, p. 333*).

A second group of studies has used average employment figures in a given industry calculated from the data for all towns as the break-point. On this basis, a town is considered as specializing in a given activity when its employment exceeds the national average. The classification of towns in the Netherlands by Steigenga (1955) is typical of studies using this approach. However, the use of national averages implies that the economy of every town is a miniature replica of the national economy. Most studies of cities have shown that the proportion of the labour force employed in a given activity varies directly with city size. An alternative, and perhaps more meaningful approach, is to use averages calculated for towns of different size classes. This method was adopted by Pownall (1953) in the classification of New Zealand towns. Figure 9.18B shows the variation in mean per cent values of total employment in the manufacturing, building and construction category (y axis) for towns of different sizes (x axis) in New Zealand. Towns for which per cent employment in this category exceed the average for its size class, and thus

fall in the shaded area in Figure 9.18B, were classified as specialized manufacturing towns.

When towns are classified in this way, no account is taken of the magnitude
of deviation above the average. Consequently, towns for which employment
is just above the average are grouped in the same category as those with
extremely large deviations. To differentiate towns taking this into account,
Nelson (1955) has suggested the use of the standard deviation. In his classification, three classes of specialization can be recognized for any given activity
depending on whether deviation is in excess of one, two or three standard
deviations above the mean respectively.

Smith (1965) points out that if the immediate purpose of classification
is to identify classes in which towns of similar functional composition are
grouped so that its members are more like each other than they are other
towns not included in the group, this will not necessarily be achieved by the
procedures used in the above-mentioned simple classifications. This is very
clearly illustrated in Figure 9.18A, where the cluster of towns enclosed by
the broken circle around the intersection of the two break-points would be
split among four groups despite the fact that each town is closer to, and is
therefore functionally more similar to, other towns in the cluster than to
other towns in the graph. At best the simple methods of classification yield
functionally heterogenous classes in which significant functional combinations may be concealed. More sophisticated taxonomic methods are needed if
classifications are to be at all meaningful. Some of the available techniques
have been reviewed by Berry (1958). Perhaps the most important attribute
of these techniques is that they permit classifications based on multiple rather
than single variables.

Typical of multi-variable classifications is that by Smith (1964) of Australian towns. Using correlation techniques on employment data for 422 towns,
he identified 91 initial groups comprising towns with similar functional
structure. Cluster analysis of the resulting correlation matrix yielded a final
classification of 17 groups. When many characteristics of cities are used in
this way as a basis for classification, it may well turn out that some of the
measures vary in similar ways from city to city and thus tell roughly the same
story about a place. When this is the case, one might suspect that a more
basic pattern underlies the variation of the different characteristics. Approximations to the basic underlying patterns may be revealed by the use of factor
analysis. This method was used in the classification of British towns by
Moser and Scott (1961). Fifty-eight measures covering eight aspects of social,
economic and demographic characteristics were included for each of the
157 towns of over 50,000 inhabitants in 1951. In the resulting classification,
towns are grouped on the basis of overall character rather than economic
specialization alone.

Variation in the original characteristics was summarized by four underlying principal components. These were identified with (1) social class structure, (2) population growth, (3) a mixture of developments after 1951 and employment structure in 1951, and (4) housing conditions. The four components accounted for about 60 per cent of the basic differences between the 157 towns. Allocation of each town to a group on the basis of 'scores' on these components yielded a classification into 14 classes each containing about 10 towns. Three main urban types emerged: (1) the resorts, administrative and commercial centres, (2) industrial towns, and (3) suburbs and 'residential' centres.

The economic base. It has been implied so far that towns cannot exist by taking in their own washing; their existence and growth depend in large measure on their ties with other areas. A town flourishes because a proportion of the goods it produces is sold beyond its borders. It follows from this that a proportion of the total labour force in any town is directly concerned with the production of goods for 'export'. These are called *basic* or city forming workers because their efforts bring money into the town, thereby enabling the purchase of raw materials, food and manufacturing goods which the town cannot produce for itself. The remaining workers can be considered as *non-basic* or city serving since their primary role in the city's economy is to service the basic sector. Consideration of this dichotomy in the labour force has long been recognized as a useful concept in the economic analysis of urban areas and, since towns can be differentiated on the basis of basic employment, as a more satisfying means of city classification. The difference in the relative importance of various industry categories when basic rather than total employment is considered is nicely revealed in Figure 9.19. In both of these towns in Wisconsin, USA, the service sector becomes less important as expected while the dominance of manufacturing in Oshkosh and government in Madison, the State capital, are clearly revealed as the major basic functions.

Crucial for the application of the economic base concept is a satisfactory method of calculating the size of basic employment. Many methods have been suggested. Alexander (1954) used the 'firm by firm' approach in which individual businesses provide information about total employment and per cent of sales to local and regional markets. Using this information, employment can be pro-rated into its basic and non-basic components. For example, a firm employing 100 workers and selling 60 per cent of its total finished product in the regional market would be assumed to have 60 basic and 40 non-basic workers. Total basic workers in a city can be obtained by summing the figures over all firms. The obvious disadvantage to the use of this method is that although it may be alright in small towns like Madison and Oshkosh shown in Figure 9.19, it becomes virtually impossible to apply to large urban areas.

A more general approach was used by Alexandersson (1956) in a study of the industrial structure of 864 cities with more than 10,000 inhabitants in the United States. The per cent of the city's total employment was calculated for each of 36 industrial categories. For each category, towns were arrayed from smallest to largest by per cent employment in a cumulative distribution. Then, for a given industry, the value of the fifth percentile which corresponded to the value for the 43rd town in the array, was arbitrarily selected as the minimum per cent employment needed to serve the needs of the city's own population. This was designated the K-value for that industry. The sum of

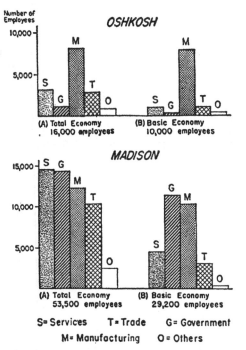

9.19 Differences in total and basic employment (*Source: Alexander, 1954, p. 250*).

all K-values for the 36 industry groups yielded the total non-basic employment in any city which was subtracted from the total employed labour force to give the number of basic workers. However, Alexandersson's interpretation of the K-values is questionable because of the weight given to small cities in the analysis; over three quarters of the towns included were smaller than 50,000 inhabitants. Morrissett (1958) in particular has provided insights into the variation of the K-values with city size and also shows that there are marked differences between regions within the United States.

A reliable short-cut method for estimating the non-basic employment in a city, which partly overcomes the objections to Alexandersson's technique, has been proposed by Ullman and Dacey (1960). Cities in the United States were divided on the basis of population size into six classes. A random sample of 38 cities was taken from each size class except from the largest class comprising places with more than one million inhabitants, from which only 14 cities were selected. For each city, the per cent of the labour force employed in each of 14 census categories was calculated. Then for each size class the figure for the city with the *minimum* per cent employed in each of the 14 categories was entered in a table. The sum of the minima for towns in a given size class was interpreted as the non-basic employment needed for cities of that size. As might be expected, non-basic employment in a given category

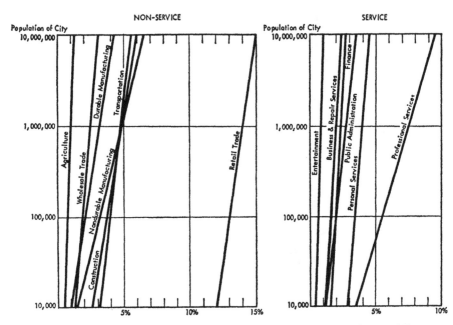

9.20 Regression lines for minimum-requirements by city size (*Source: Ullman and Dacey, 1962, p. 129*).

is directly related to the size of the city. Moreover, as Figure 9.20 shows, when the minima are plotted on a semi-logarithmic graph, straight line relationships are obtained. Minimum per cent employment figures for a town of any population size can be read off the graphs for each of the 14 categories, and summed to give the total non-basic employment. Basic employment can then be determined by subtracting this percentage from the total employment for the city.

The methods used to identify the basic employment of cities should be used with caution. Isard (1960) has listed some of the difficulties encountered in the use of employment data for this purpose and Roterus and Calef (1955) have stressed the effect that the size of area used as a base for measuring basic employment has on results. The strength of economic base models is their simplicity; it might however appear that simplicity is an attribute of declining importance in geographical studies. If the methods of determining basic employment become too complex, for example see recent methods proposed by Hoyt (1961), the appeal of economic base studies is largely lost and the concept must compete on unfavourable terms with other analytical tools like inter-industry linkage analysis (Isard and Kavesh, 1954) which are capable of providing more elegant results.

THE INTERNAL STRUCTURE OF CITIES

Models of urban structure are basically of two kinds: (1) *partial*, which deal with the location of a specified set of activities (e.g. residential land use) based on assumptions about the locational characteristics of all other activities in the urban area, and (2) *comprehensive*, which deal with the location of all activities together within the urban area. In both types of models the importance of transport in determining land use patterns is stressed, either explicitly in terms of the substitution of rents for transport costs, or implicitly in terms of land value-accessibility relationships which Hoyt (1939A) thought of as a common denominator of all land uses.

The urban land market

Land use patterns result from a multitude of decisions made by individuals about location. It is not at all clear how these decisions are reached although recent studies using gaming-simulation techniques are beginning to shed some light on the underlying mechanisms involved (Duke, 1964). But no matter what the underlying considerations are, it appears that decisions are regulated in varying ways by the economic processes operating in society. Ratcliff (1949), elaborating the earlier ideas of Hurd (1911) summarises this when he says that '. . . the locational patterns of land use in urban areas result from basic economic forces, and the arrangements of activities at strategic points on the web of transportation is a part of the economic mechanism of society'.

The pertinent aspects of this mechanism as it relates to the generation of land use patterns can be briefly summarized as follows. Each activity has an ability to derive utility from every site in the urban area; the utility of a site is measured by the rent the activity is willing to pay for the use of the site. The greater the derivable utility, the greater the rent an activity is willing to pay. In the long run, competition in the urban land market for the use of available sites results in the occupation of each site by the 'highest and best' use, which is the use able to derive the greatest utility from the site and which is, therefore, willing to pay most to occupy it. As an outgrowth of the occupation of sites by 'highest and best' uses, an orderly pattern of land uses results in which rents throughout the systems are maximized and all activities are optimally located. This process is, of course, identical with the original formulation by von Thünen (1826), and later modifications by Dunn (1955), of the ordering of land uses in agricultural regions in relation to market centres. However, as Alonso (1960) and Wingo (1961) have pointed out, the operation of the urban land market is rather more complex than its agricultural counterpart.

The rent paid for the use of a site is affected by many factors, but most importantly by the location of the site relative to other uses. The logic of this relationship is founded on the assumption that site rents represent a saving in transport costs in overcoming the 'friction' of distance. From this it is argued that competition for the use of land results in the minimization of the 'friction' of distance in the entire urban area and, since accessibility increases inversely with distance, the resulting pattern of urban rents is essentially a function of transport. Savings in transport costs can be traded off for extra rent payments to ensure the use of a particular site. Therefore those activities which enjoy the greatest benefits from occupying accessible locations will have greater surpluses available with which to bid for land. Consequently, sites in the urban area are not merely occupied by activities which can pay most for their use, but more specifically, by those activities which are able to derive the greatest positive transport advantages from the use of a given piece of land. When rents are viewed in this framework they will be represented by land values, which in turn can be considered as a direct reflection of differences in intra-urban accessibility. Thus, high land values will be associated with highly accessible locations and *vice versa*.

The land value surface

If land values are the common denominator of land uses, an understanding of the pattern of land values would help in understanding the internal structure of cities. It is perhaps somewhat surprising therefore that until recently little attention has been paid to this matter by geographers. The specific pattern of land values, just like its equivalent pattern of land uses, obviously varies from city to city depending upon local circumstances. At least three elements are present in the pattern in all cities, namely (1) land values reach a grand peak in the centre of the city and decrease by varying amounts outward toward the periphery of the urban area, (2) land values are higher along the major traffic arteries than in the areas away from them, and (3) local peaks of higher value than the general level at a given distance from the city centre occur at the intersection of major traffic arteries. Superimposition of these three components results in a general surface of land values similar to that shown diagrammatically in Figure 9.21. This surface can be likened to a conical hill, the smooth surface of which is disturbed by ridges, depressions and minor peaks.

Although many factors affect the value of a given piece of land, for example elevation has a noticeable effect on residential values (Brigham, 1964), we argue here that the land value surface is essentially a direct reflection of accessibility within the urban area. Accessibility is highest at the city centre because it has developed through time as the major focus of routes and consequently is the most easily reached part of the entire urban area. Competition

9.21 Generalized land value surface within a city (*Source: Berry, Tennant, Garner and Simmons, 1963, p. 14*).

for the use of land is most intense here. In a similar way, accessibility is greater at sites located along the radial and circumferential routes and at the intersections of these than it is away from them. Moreover, since some parts of the urban area are better served with transport than others, the value surface will not be characterized by uniform slope in every direction but will decline more rapidly with distance from the central peak in some directions than in others. The result is a marked sectoral variation in the general level of the value surface.

Accessibility means different things for different activities. For commercial functions, proximity to sales potential would appear to be crucial for retail activities while proximity to complementary uses is critical for offices requiring face-to-face contacts. For the variety of industrial uses, accessibility is in terms of availability of suitable land, transport facilities, public utilities and, in the case of agglomerations, proximity to other industries with which there are technological linkages. In all these instances, accessibility relates directly to costs of operation and profit levels. This is not so in the case of the third general land use type, residential activity. Although proximity to workplaces, shopping centres and other amenities is important, decisions about residential location are made within the framework of the more intangible environmental qualities that Alonso (1960) argues underlie 'satisfaction' for the household.

Following the classic work of Hoyt (1939A) in Chicago, the findings of other empirical studies support the general nature of the land value surface. Typical of these studies is that by Seyfried (1963) in which directional differences in the exponential slope of land value profiles were clearly identified in Seattle, USA. Significant correlations ($R^2 = 0.64$) were obtained between land values and distance away from the centre of the city but comparison of the slopes of regression lines for sectors going north, south, east and west from the city centre indicated the western profile was nearly three times as steep as that for the northern sector, while the profiles for the other two sectors were intermediate in slope. Similar evidence is given by Knoss (1962) for Topeka, Kansas, USA, where values were found to decline inversely with the reciprocal of distance from the city centre and from major radial routes, while they appeared to vary directly with the direction of growth within the city. Sectoral variation in land values has been identified in Chicago by Hayes (1957) and Mayer (1942), and more recently Berry, Tennant, Garner and Simmons (1963) have indicated the relationship between local peakings on the value surface and the relative importance of highway intersections.

The most comprehensive study to date is that by Yeates (1965) who hypothesized six variables in a multiple regression model to explain variation in the value surface in the city of Chicago. These were namely, (a) distance from the central business district, (b) distance from the nearest regional level shopping centre, (c) distance from Lake Michigan, (d) distance from the nearest elevated-subway line, (e) per cent non-white population of the block in which a particular site is located, and (f) population density. The model provided a low degree of explanation ($R^2 = 0.18$) of variation in the surface when applied to data for 1960. However, when a sectoral component was added, about half of the total variation ($R^2 = 0.51$) was accounted for, which attests the importance of radial routes in determining land values.

Land use models

The internal structure of any city is unique in its particular combination of detail. In spite of this, it appears that in general there is a degree of order underlying the land use patterns of individual cities. Since there is as yet little agreement on the specific nature of this order, a variety of models have been proposed. In line with the three basic components identified in the land value surface, they will be discussed here under concentric, sector and nucleii headings.

Concentric models. Models of this kind are developed on the assumption that land values, and by implication accessibility, decline with equal regularity in all directions from a common central point in the city. Since no account is taken of distortions caused by differential accessibility, land use patterns are assumed to be arranged in regular concentric zones about the city centre.

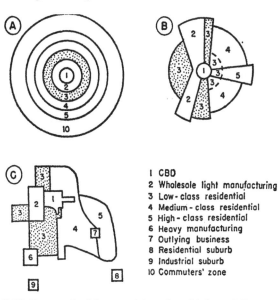

1 CBD
2 Wholesale light manufacturing
3 Low-class residential
4 Medium-class residential
5 High-class residential
6 Heavy manufacturing
7 Outlying business
8 Residential suburb
9 Industrial suburb
10 Commuters' zone

9.22 Concentric (A) sector (B), and multiple nucleii (c) models of urban structure (*Source: Haggett, 1965, p. 178*)

The classic concentric zone model (Fig. 9.22A) was proposed by Burgess (1925) based largely on his studies of the Chicago region. The model states that at any given moment in time land uses within the city are organized into zones differing in age and character and located in a definite order from the city centre. For Chicago, five zones were identified; these were in order from the centre outwards: (1) an inner central zone which formed the 'heart' of the city's commercial, social, cultural and industrial life and which was the focus of urban transport, surrounded by (2) a transition zone of mixed land uses in which deteriorating residential property predominates. Its unattractiveness is emphasized by the blighted conditions and slums intermixed with light industries and other business uses which have spilled over from the inner core; (3) a working class residential zone in which second generation immigrants form an important element in the population structure; (4) a zone of better housing characterized by single-family dwellings interspersed with pockets of exclusive residences and high class apartment buildings; and (5) a fringe zone of suburban and satellite communities forming dormitory suburbs for people working in the central city.

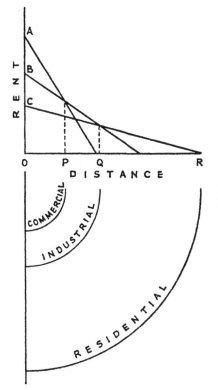

9.23 Hypothetical rent-distance relationships within a city (*Source: Garrison*, et al., *1959*, *p. 64*).

Although no explanations were offered by Burgess for this particular arrangement of land uses, Berry (in Garrison *et al* 1959) and Isard (1956, p. 200) provide a partial understanding in terms of the substitution of rents for transport costs. Accessibility is assumed to decrease uniformly in all directions from the city centre, and by implication the rent that an activity is prepared to pay for site use decreases from the central point. The resulting rent curves will slope downward to the right, as shown for example, for activity A in Figure 9.23. Not all activities are equally susceptible to differences in accessibility and their rent curves will vary in steepness accordingly, as shown for activities B and C in Figure 9.23. Since the rent curves for activities

A, *B* and *C* are ranked in steepness, they are also ranked in order of accessibility from the city centre. In competition for the use of sites, activity *A* outbids all other uses and occupies locations from the centre of the city (*O*) to distance (*P*); activity *B* is able to pay the highest rent and occupies sites between (*P*) and (*Q*); while activity *C* is only successful in bidding for sites between (*Q*) and the edge of the city (*R*). When this arrangement is generalized from one to two dimensions by rotation of (*OR*) about (*O*), a simple concentric pattern of land uses is obtained and if activities *A*, *B*, and *C* are imagined as commercial, industrial, and residential uses respectively, the result is a simplified version of the concentric zone model.

One of the interesting features of the concentric model is the anomaly of poor people living close to the city centre on high value land and the rich living at the periphery where land is cheaper. Both Alonso (1960) and Wingo (1961) have partially explained this in models of transport and urban land use. The specific details of the economic arguments proposed are too complex to be elaborated here although the general argument is simplified as follows. Each household has a limited budget with which it must satisfy all its needs, viewed essentially as (a) basic costs of living, (b) costs of housing, and (c) costs of getting to and from the centre of the city where it is assumed all workplaces are concentrated. Once the basic living costs of the household are established, the remaining funds can be allocated in varying proportions between housing and commuting costs. Poorer families, with less money available for commuting live close to workplaces but can only afford to pay for small amounts of the high value land near the city centre. Conversely, richer households can afford to live more luxuriously at the edge of the city where they consume larger amounts of lower value land at the expense of paying higher commuting costs to the city centre.

Although Blumenfeld (1949) claims that the concentric pattern is found in Philadelphia and Smith (1962) has recognized some of the zones in Calgary, Canada, distortions will inevitably be introduced by natural barriers and the pattern of transport routes. Hartman (1950) shows that the circular shaped city form necessary for the development of pure concentric land use patterns will only result if there is a strong radial component in the transport system, and then only if the number of radial routes is quite large and they are closely spaced. Fewer radial routes spaced more widely apart give rise to marked differences in intra-urban accessibility to result in a 'star' shaped form in which concentric arrangements of land use are distorted and even destroyed.

Sector Models. Models of this type are developed on the assumption that the internal structure of the city is conditioned by the disposition of routes radiating outwards from the city centre. Differences in accessibility between radials causes marked sectoral variation in the land value surface and correspondingly an arrangement of land uses in sectors. This is the basis of the model proposed by Hoyt (1939B) who hypothesizes that similar land uses concentrate along a particular radial route from the city centre to form sectors as shown in Figure 9.22B. Thus a high rent residential district in one sector of the city would migrate outwards in that direction by the addition of new growth on its outer arc. Similarly, low rent districts in another part of the city would develop by the same process in that direction. The sectors of high-class residential areas would seem to be particularly pronounced in the direction of high ground and open spaces.

Although the model is descriptive, it is clearly an improvement on the earlier concentric zone idea since both distance and direction from the city centre are taken into consideration. Perhaps because of this, sectoral arrangements have been identified in many cities. There is evidence of a sector of high-class residential land use with high land values along the northern shore of Lake Michigan and sectors of low-class residences in the south and industry in the west of Chicago (Yeates, 1965). For Belfast, Jones (1960) found that the pattern of high-class residential areas was consistent with the sector model, while Smith (1962) claims that sectors rather than concentric zones seem to be most meaningful in Calgary, Canada. Although approximations to an inner core area surrounded by a transition zone and an outer commuting zone are found in this city, the two middle residential zones postulated in the concentric model are not apparent in any form. Most of the major land uses however, formed something like sectors.

Nucleii models. Land use patterns in most cities are not built around the single centre postulated in the above models; rather they are developed around several discrete centres within the urban area. This is the basis of the multiple nucleii model of Harris and Ullman (1945) shown in Figure 9.22C. The number and location of the nucleii within the urban area depends on the size of the city, its overall structure and historical development. The larger cities have a greater number and more specialized nucleii than smaller places. For American cities, five districts were identified: (1) the central business district, (2) a wholesaling and light-manufacturing area near the focus of extra-city transport facilities, (3) a heavy industrial district near the present or former outer edge of the city, (4) various different residential districts and (5) peripheral dormitory suburbs. The reasons given for the existence of separate nucleii and differentiated districts were combinations of (1) the specialized requirements of certain activities, (2) the tendency for activities to agglomerate, (3) the repulsion of some activities by others which is linked to (4) the differences in rent paying ability which force activities to cluster in separate districts within the city.

One of the many implications of this model is that marked differences in type of residential land use should be noted around the various business nucleii in the city. Marble (in Garrison *et al.* 1959) has proposed a series of models in which residential site selection, measured by land values, is hypothesized as a function of location within the city in respect to its centre and other major business nucleii. Application of the models to randomly selected city blocks in Spokane (Washington) and Cedar Rapids (Iowa), USA, yielded non-significant results. It was concluded that the distribution of residential land values, and by implication the type of residential property, provides no support for the concept of multiple nucleii or even concentricity in residential pattern.

The three models are not mutually exclusive and elements of all three patterns might be expected in cities, especially where they have fused with one another to form the large conurbation areas or Megalopolis (Gottmann, 1961), the urbanized north-eastern part of the United States, with more complex internal land use patterns. Marble (in Garrison et al. 1959) has suggested a fused model in which growth proceeds radially from the city centre and from various other nucleii but is intercepted by axial growth pushing outwards along the lines of least resistance from the main centre to result in a star shaped city in which distinct social, economic and technical zones are developed. Haggett (1965, p. 180) offers evidence from an area south of Cambridge, England, suggesting that Marble's views may be more appropriate. In this area, the spread of housing since World War II shows distinct gradients close to the city, along major radials and in the outlying villages which were identified with the concentric, sector and nucleii models respectively. Clearly, all three models are useful in generalizing about the internal structure of cities and it might well be that judicious application of analytical tools like factor analysis will enable a more detailed description of the underlying components of city structure in terms of these simple models and enable the formulation of even more general and meaningful models to supplement them.

Urban population densities

Regardless of the particular land use arrangement within cities, the distribution of population densities appears to be the same in all cities. Densities decline as a negative exponential function of distance from the city centre. This can be generalized as

$$P_d = P_o e^{-gd}$$

in which P_d is the population density at a given distance (d) from a point in the central business district; ($-g$) is the slope of the density decline curve, and P_o is the density of the central area (extrapolated from the slope for outer areas). The regularity, shown for Hyderabad in India and Chicago, USA, in Figure 9.24, was originally identified by Clark (1951) from a study of 36 cities in various parts of the world, and many of the subsequent studies of urban densities are summarized by Berry, Simmons and Tennant (1963).

Although to date no city has been studied for which the regularity does not hold, marked variation in the slope of the gradient is characteristic. Muth (1962), for example, has shown that the g-values for 36 American cities ranged between 0·7 and 1·2, although the majority of values clustered in the range 0·2 to 0·5. Attempts to explain these variations led Muth to carry out a regression study of density gradients against twelve controlling variables of which only four (car registration per capita in the urban area, the proportion of manufacturing employment and of substandard dwellings in the central area, and the size of the urban area) were significant in accounting for differences in slope of the curves. This led to the conclusion that larger cities with low transport costs, dilapidated central parts and dispersed employment centres were more compact than other cities. However, a study by Berry

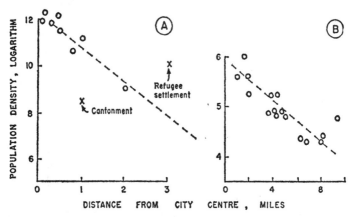

9.24 Population density-distance relationships for Hyderabad, India (A) and Chicago, USA (B) (*Source: Haggett, 1965, p. 157. After Berry, Simmons and Tennant, 1963, pp. 392 and 394*).

(Berry *et al.*, 1963) of 46 American cities suggests that the spatial pattern of manufacturing activity is not one of the factors associated with variation in the slope of density gradients. Variation also typifies central densities, which Winsborough (1962) has shown to be a function of overall population density within the city, which in turn is directly related to population size of the town, its importance as a manufacturing centre and the proportion of old dwellings in the city. Added evidence of the relationship between population densities and age of cities is given by Berry (Berry *et al.*, 1963) who found that a regression equation with age of city and the slope of the density gradient as independent variables could account for approximately sixty per cent of the variation in the size of central densities in his sample cities. The implication from these studies is that larger, older industrial cities tend to have higher overall population densities than other kinds of cities.

Although there is no complete explanation for the empirical regularity, recent works by Alonso (1964) and Muth (1962) suggest that it results from substitution on the part of households of rents for transport costs as discussed above, and confirm Clark's (1951) original speculation that the observed regularity had something to do with transport costs. If the rent curves for lower income groups is steeper for any pair of households with identical tastes, the poor will live at high densities near the city centre while the rich will live at lower densities near the periphery. The reason for the negative exponential nature of the density-decline function is claimed by Muth (1962) to stem from the similar form of the function for housing production. Both these sets of findings suggest that the negative exponential equation is a general one which can be derived as a logical extension of the theory of the urban land market. Moreover, when placed in the framework of general systems theory, Berry (1964) shows that the pattern of population densities within cities can be fairly accurately predicted if the population size and age of city are known, for the density gradient (g) and central densities (P_0) are related to the position of the city in, and the slope of, the rank-size curve. It seems that what were considered hitherto as disparate empirical regularities may be synthesized as part of a wider model of city structure.

Commercial structure. Apart from references to the central business district, the general models of internal structure do not say much about the nature of the urban business complex. A descriptive model of this complex is given by Berry (1959) for American cities, and the major components recognized are shown in Figure 9.25. This description is the result of analysis of locational

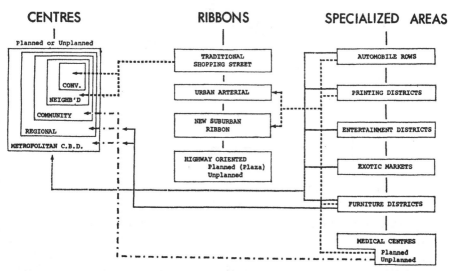

9.25 Elements of the commercial structure of American cities (*Source: Berry, Tennant, Garner and Simmons, 1963, p. 20*).

requirements and spatial associations of different business functions and is unlike earlier studies of commercial structure, for example the works of Mayer (1942) or Kelly (1956), which stressed differences in morphological structure rather than the functions typical of the different conformations.

Three basic conformations are recognized: (1) older unplanned and newer planned shopping centres, which provide a variety of convenience and shopping goods from functionally linked establishments for essentially home-based shopping trips of various frequency and duration; (2) a variety of ribbon developments which, with the exception of the traditional shopping streets, consist of a range of space consuming services catering primarily to demands originating on highways, and (3) specialized areas catering to special consumer demands. Differences within each of the major types are related to the numbers and kinds of functions performed and a detailed discussion of their functional structure is provided in Berry, Tennant, Garner, and Simmons (1963).

Although there are studies of various aspects of ribbons, for example Faithfull (1959) in Australia, Foster and Nelson (1958) in Los Angeles, Berry (1959) in part of Washington State, USA, and Boal and Johnson

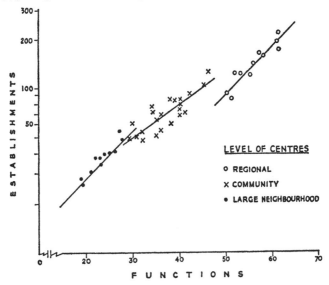

9.26 Three orders in the hierarchy of shopping centres in the city of Chicago, USA (*Source: Garner, 1966*).

(1965) in Canada, most emphasis has been given to the study of nucleated centres, and especially the central business district in studies of commercial structure. Shopping centres can be viewed as the urban equivalent of rural central places and the basic notions of central place theory can be applied to their study. Analysis of functional structure, size and trade area characteristics shows that, like central places in rural areas, there is a hierarchy of business centres within the city. This has been recognized by many researchers, for example by Smailes and Hartley (1961) and Carrothers (1962) in Greater London, and Thorpe and Rhodes (1966) in the Tyneside urban region, in

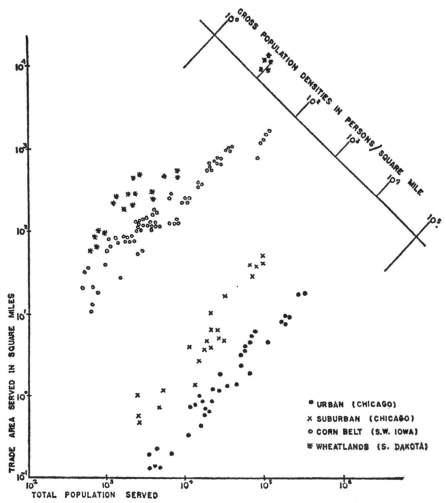

9.27 Hierarchies under different density conditions in the United States (*Source: Berry, and Barnum, 1962, p. 40*).

which subjectively selected criteria were used in identification and delimitation. Other studies have used more rigorous analytical techniques, for example Berry, Tennant, Garner, and Simmons (1963) used factor analysis in a study of Chicago, and Garner (1966) used regression and covariance analysis to identify the hierarchy in the city of Chicago. For this city these studies have identified three levels below the central business district. In ascending order, these are: (1) *neighbourhood* centres offering convenience goods for people living locally, (2) *community* centres providing infrequently demanded goods for several neighbourhoods, and (3) *regional* centres supplying specialized goods for people living in a major portion of the urban area. The number of levels in the hierarchy depends very much on the size of the central business district, which in small towns may be the equivalent of a regional centre for a large metropolitan area like Chicago. A hierarchy for the city of Chicago is illustrated in Figure 9.26 from the relationship between number of functions (x axis) and number of shops (y axis) in 68 shopping centres. The average regional, community and neighbourhood centre provide 56, 37 and 24 functions respectively. Comparison of these sizes with those shown in Figure 9.15A for cities, towns and villages in rural Iowa, USA, show a surprising degree of consistency in functional size, and suggests that intra-urban hierarchies may be viewed as a logical part of a series of hierarchies developed under different population density conditions (Berry and Barnum, 1962) as shown in Figure 9.27.

More detailed analysis of the hierarchy of shopping centres in Chicago by Berry (Berry *et al*, 1963) suggests the existence of more than one hierarchy within cities. The overall relationship between number of functions and shops is shown in Figure 9.28 to mask noticeable differences in the relationship for various parts of the city on the one hand and between the older and newer centres on the other. The regime in the poorer part of the city is of a different order to that in the high income area, and the lower correlation between numbers of functions and shops ($R^2 = 0.69$) suggests greater duplication of functional types in areas where demand is lower and of a different nature. Both Pred (1963) and Garner and Harvey (1965) have provided evidence of the different functional composition and quality of establishments in shopping centres serving lower income groups in the city. Planned centres appear to provide fewer functions less frequently than their unplanned counterparts owing to the greater selectivity in the design of centre size and form.

Central place theory does however have limitations for studying internal business structure; without considerable mental gymnastics it is difficult to apply to the study of urban business ribbons. This shortcoming has prompted Curry (1962) to suggest that a completely new approach to central place studies of the internal business structure of cities is required and he attempts to provide the framework of an operational approach in a probabilistic formulation of consumer behaviour using the poisson series.

A definite relationship appears to exist between the location of the various components of urban business structure and land values. Ribbons are associated with the ridges of higher land values along major traffic arteries while centres are related to the local peakings in value at highway intersections.

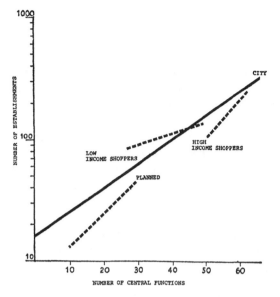

9.28 Different relationships of shops (y-axis) to number of functions (x-axis) for Chicago, USA (*Source: Berry, Tennant, Garner and Simmons, 1963, p. 134*).

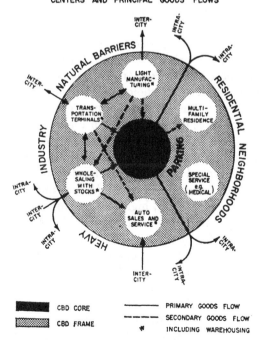

9.29 Diagrammatical structure of the central business district (*Source: Horwood and Boyce, 1959, p. 21*).

Stratification in value peaking for the latter might be expected to correspond to stratification of centres within the hierarchy and there is some general evidence for this in Chicago (Berry et al., 1963). Here the smaller neighbourhood centres are generally found at places with values about fifty dollars above the surrounding general level, while regional centres, needing more accessible locations to serve larger trade areas, are found where peaks rise about 4,000 dollars above the general level. However, such factors as stage of growth of centres, distance from the central business district, quality of market area served and lag in values accompanying decline distort the regularity and prevent the use of peak values as a proxy variable in the identification of centres (Garner, 1966).

The relationship between land values and location of commercial activities is more clearly apparent in the arrangement of functions *within* shopping centres. Many studies have shown this for the central business district, for example in Glasgow's central area studied by Diamond (1962), and it can be viewed as one of the explanatory variables underlying the 'core-frame' model of central business district structure proposed by Horwood and Boyce (1959). They argue that the commercial centre comprises two distinct parts, as shown in Figure 9.29: (1) an inner core of intensive use of high value land resulting in marked vertical expansion, where strong functional links between shops and between various offices is reflected in clusters of functions forming distinct micro-land use zones, and (2) a less intensively developed frame where values are relatively lower and functions have very little in common except location.

The relationship between land uses and functional arrangement is more clearly specified for centres below the central business district in the hierarchy in a model proposed by Garner (1966). He suggests that competition between functions of different threshold sizes for the use of sites leads to an ordered arrangement of land use within shopping centres, the salient features of which are shown for three levels of centre in Figure 9.30. For a given level of centre, the inner core area of high value land is occupied by functions which set the centre above the level of others in the hierarchy and is surrounded by functions typical of each lower level of centre on successively lower value land. Thus, in a regional centre (Fig. 9.30A) the core comprises high threshold, and as it turns out, high rent paying regional level functions surrounded in turn by community level and neighbourhood level functions. At the community level (Fig. 9.30B) regional types are excluded by definition and the core area is pre-empted by community functions which are surrounded on lower value land by neighbourhood types. Neighbourhood centres (Fig. 9.30C) have a simple structure since only one level of function is provided. Tests of the model in the city of Chicago show that the hypothesized patterns agree with the general structure of centres but that at the regional level, the model tends to oversimplify the real world conditions.

Before closer agreement between model and reality is to be obtained, refinements are needed in classification of business types and a more thorough appraisal of the threshold concept and product differentiation between establishments must be developed.

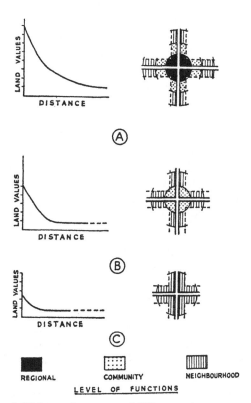

9.30 Internal structure of different order shopping centres in cities (*Source: Garner, 1966*).

Growth and change in internal structure. The models discussed so far represent cross-sections of the urban scene at a given moment of time. But the city is dynamic and is constantly in a state of flux accompanying growth of its constituent parts, either by extension at the periphery of the urban area or by internal re-arrangement of existing land uses. Both result in changes in intraurban accessibility and attendant changes in the pattern of land values, which in turn is reflected in the changing pattern of locations. Yeates (1965) has amply demonstrated this in Chicago where his model of land values (discussed above) accounted for only 18 per cent of variation in the surface for 1960 but accounted for over 75 per cent of the total variation in 1910. Changes in mobility associated with increasing use of the automobile and continued decentralization of population has caused the effects of distance from the central business district and the influence of rapid transit routes to be less important and an increase in values along new radial and circumferential routes. Growth in the normative models discussed above is accounted for by the ecological concept of invasion-succession. This states that occupancy of land by new uses tends to make it unsuitable for further occupancy by the original tenants who move out to new locations. In the concentric zone model new locations for functions of one zone are found in the next outer adjacent zone, so that growth consists essentially of colonization outwards on a broad front. In this way, the innermost zones are subject to continued decay and outer zones by gradual modification as lower status occupants move in. In a similar way, the innermost part of a given sector of land use changes in character as new uses take over and the sector pushes out by accretion on its outer arc.

Empirical studies of invasion-succession are abundant in the literature of Human Ecology, for example see Hawley (1950). Recently, Morrill (1965) has provided a more detailed understanding of the processes as they pertain to the extension of negro areas in American cities using Monte Carlo simulation techniques, and Hoyt (1964) has specified the factors responsible for distorting the classical models of urban structure. According to Smith (1962) it appears that these result in a progressive breakdown of the rigid concentric and sectoral zonations through time causing patterns to conform increasingly to the more general arrangement hypothesized in the multiple nucleii model.

Concomitant with reorganization of the internal land use patterns, growth by extension of the built-up area takes place at the periphery. Chapin and Weiss (1962) provide some insight into the factors influencing the development of urban land using multiple regression models in a study of growth in a cluster of towns in North Carolina, USA. They conclude that development is discouraged by poor drainage and proximity to blighted and non-white areas. Conversely, growth is intensified in areas with good transport facilities, near to large employment centres and in fringe settlements provided with a wide range of community services.

An attempt to specify the relationship between population growth and the characteristics of suburbs has been made by Thomas (1960) for the urbanized area of Chicago in the decade 1940–50. He hypothesized that population growth in suburbs at the periphery of the built-up area is a function of population size and density (a proxy variable for vacant land) of the place, age of suburb, cost of housing, quality of schools, birth-death differentials, degree of industrialization and accessibility to the central city. However, these only accounted for about a third ($R^2 = 0.36$) of variation in population growth in the entire study area, and birth-death differential and population density alone were significant. The model was more successful when fitted separately to three sectors within the city and presumably illustrates differences in residential desirability within the urban area.

Other studies have investigated the way in which growth takes place at the edge of the city. Blumenfeld (1959) has formulated a model in which growth takes place in the form of waves at the edge of the city although studies by Garner (1960) tend to indicate that growth 'leapfrogs' backwards and forwards between suburbs with no discernible regularity. Something similar to a wavelike expansion exists in relation to the disposition of transport routes. He suggests a model in which growth takes place first at locations on radial

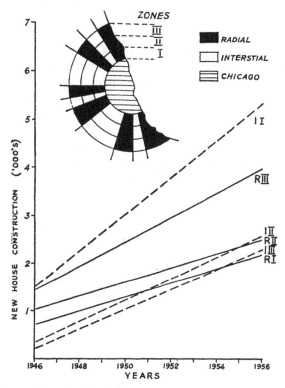

9.31 Regression lines for residential expansion in the Chicago urbanized area, USA (*Source: Garner, 1960, p. 30*).

routes and occurs at a later period in the interstitial areas between them. Evidence in support of this hypothesis is presented in Figure 9.31 which shows trends in new house construction at radial and interstitial suburbs at various distances from the central city during the period 1946–56, in the Chicago urbanized area. The number of new houses constructed in zone one at the edge of the city (10–15 miles from the centre) is much greater at interstitial suburbs (I_1) than at places on radial routes (R_1) for that distance. In zone II, building continued to push out along radial routes (R_2) but by 1956 this growth had been surpassed by residential growth at interstitial suburbs (I_2). In the outermost zone III, growth is predominantly at radial sites (R_3) with relatively little new house construction at corresponding interstitial locations (I_3). The implications from this are: (1) growth occurs at different rates in sectors which have good accessibility with the central city, and (2) expansion does not take place along a broad radial front as posited in the concentric zone model but is more in line with the general ideas of the sector model although the lag in development of interstitial areas is noticeable.

The continued outward expansion of the urban area and redistribution of population seems to be reflected in changes in the population density gradients. Berry, Barnum, and Tennant (1963) have shown that the gradient tends to flatten out accompanying low density extension of the periphery and a lowering of the high central densities as people spread out into adjoining residential zones. These changes are in turn related to modification of the commercial pattern. As areas are invaded by lower income groups, changes in the nature of demands result in the reduction of numbers of shops and functions provided, and perhaps more importantly, in the kinds of functions offered as blight conditions set in. At the same time, new shopping centres grow up to serve the increasing demands in outlying districts, shopping habits change and older centres decline (Simmons, 1964). The central business district becomes less important and sales decline in face of competition from outlying shopping centres (Boyce and Clarke, 1963). This in turn is associated with a shrinkage of the core area and the gradual encroachment of the frame uses to give rise to a 'zone of discard' which Murphy and Vance (1955) have shown to be characterized by the invasion of low grade establishments.

CONCLUSION

We have illustrated some of the ways in which models are being used in the study of settlement location and urban geography. The coverage is by no means complete; the various alternative approaches proposed by sociologists, econometricians, and the recent wealth of studies of urban structure provided by traffic engineers and planners have been passed over. Three remarks may be in order by way of summary: (1) Urban research is still in a relatively primitive stage of development and in spite of the voluminous literature much more work is needed to provide a comprehensive understanding of the complex integrated growth and structure of urban areas. (2) The study of settlement by geographers must progress hand in hand with workers in allied fields if spatial models are to be at all meaningful. (3) The paucity of references in the chapter to the use of models in British studies speaks for itself; there is a very obvious need for the development of models to be used in research and teaching if geographers are to contribute to the study and understanding of the spatial structure of urban areas.

REFERENCES

ALEXANDER, J. W., [1954], The basic-nonbasic concept of urban economic functions; *Economic Geography*, 30, 246–261.

ALEXANDERSSON, G., [1956], *The Industrial structure of American cities*, (Lincoln, Nebraska).

ALONSO, W., [1960], A theory of the urban land market; *Regional Science Association, Papers and Proceedings*, 6, 149–157.

ALONSO, W., [1964], *Location and Land use*, (Cambridge, USA).

BECKMAN, M. J., [1958], City hierarchies and the distribution of city size; *Economic Development and Cultural Change*, 6, 243–248.

BERRY, B. J. L., [1958], A note concerning methods of classification; *Annals of the Association of American Geographers*, 48, 300–303.

BERRY, B. J. L., [1959], Ribbon developments in the urban business pattern; *Annals of the Association of American Geographers*, 49, 145–155.

BERRY, B. J. L., [1960], The impact of expanding metropolitan communities upon the central place hierarchy; *Annals of the Association of American Geographers*, 50, 112–116.

BERRY, B. J. L., [1961], City size distributions and economic development; *Economic Development and Cultural Change*, 9, 573–588.

BERRY, B. J. L., [1964], Cities as systems within systems of cities; *Regional Science Association Papers and Proceedings*, 13, 147–163.

BERRY, B. J. L. and BARNUM, H. G., [1962], Aggregate relations and elemental components of central place systems; *Journal of Regional Science*, 4, 35–68.

BERRY, B. J. L., BARNUM, H. G. and TENNANT, R. J., [1962], Retail location and consumer behaviour; *Regional Science Association, Papers and Proceedings*, 9, 65–106.

BERRY, B. J. L. and GARRISON, W., [1958A], Functional bases of the central place hierarchy; *Economic Geography*, 34, 145–154.

BERRY, B. J. L. and GARRISON, W., [1958B], Recent developments of central place theory; *Regional Science Association, Papers and Proceedings*, 4, 107–120.

BERRY, B. J. L. and GARRISON, W., [1958C], A note on central place theory and the range of a good; *Economic Geography*, 34, 304–311.

BERRY, B. J. L. and GARRISON, W., [1958D], Alternate explanations of urban rank-size relationships; *Annals of the Association of American Geographers*, 48, 83–91.

BERRY, B. J. L. and MAYER, H. M., [1962], Comparative studies of central place systems; *Office of Naval Research, Contract NONR 2121-18*.

BERRY, B. J. L. and PRED, A., [1961]. Central place studies: a bibliography of theory and applications; *Regional Science Research Institute, Bibliography Series*, 1.

BERRY, B. J. L., SIMMONS, J. W. and TENNANT, R. J., [1963], Urban population densities: structure and change; *Geographical Review*, 53, 389–405.

BERRY, B. J. L., TENNANT, R. J., GARNER, B. J. and SIMMONS, J. W., [1963], Commercial structure and commercial blight; *University of Chicago, Department of Geography, Research Paper*, 85.

BLOME, D. A., [1963], A map transformation of the time-distance relationships in the Lansing Tri-County area; *Michigan State University, Institute for Community Development and Services*, (East Lansing, Michigan).

BLUMENFELD, H., [1949], On the concentric circle theory of urban growth; *Land Economics*, 25, 209–212.

BLUMENFELD, H., [1959], The tidal wave of metropolitan expansion; *Journal of the American Institute of Planners*, 25, 3–14.

BOAL, F. W. and JOHNSON, D. B., [1965], The functions of retail and service establishments; *Canadian Geographer*, 9, 154–169.

BOGUE, D. J., [1949], *The structure of the metropolitan community: a study of dominance and subdominance*, (Ann Arbor, Mich.).

BOYCE, R. and CLARK, W. A. V., [1963], Selected spatial variables and central business district retail sales; *Regional Science Association, Papers and Proceedings*, 11, 167–193.

BOYCE, R. and CLARK, W. A. V., [1964], The concept of shape in geography; *Geographical Review*, 54, 561–572.

BRACEY, H. E., [1962], English central villages: identification, distribution and functions; *Lund Studies in Geography, Series B, Human Geography*, 24, 169–190.

BRIGHAM, E. F., [1964], A model of residential land values; *RAND Corporation*, (Santa Monica, California).

BRUNHES, J., [1925], *La Géographie Humaine*, (Paris).

BRUSH, J. E., [1953], The hierarchy of central places in southwestern Wisconsin; *Geographical Review*, 43, 380–402.

BRUSH, J. E. and BRACEY, H. E., [1955], Rural service centres in southwestern Wisconsin and southern England; *Geographical Review*, 45, 559–569.

BUNGE, W., [1962], Theoretical Geography; *Lund Studies in Geography, Series C, General and Mathematical Geography*, 1.

BURGESS, E. W., [1925], The growth of the city; *The City*, 47–62, (Chicago).

CARROTHERS, W. J., [1962], Service centres in Greater London; *The Town Planning Review*, 33, 5–31.

CHAPIN, F. S. JR., and WEISS, S. F. (Editors) [1962], *Urban Growth Dynamics*, (New York).

CHRISTALLER, W., [1933], *Die zentralen Orte in Süddeutschland: Eine ökonomisch-geographische Untersuchung über die Gesetzmässigkeit der Verbreitung und Entwicklung der Siedlungen mit städtischen Funktionen*, (Jena).

CHRISTALLER, W., [1966], *Central Places in Southern Germany*, (Trans. C .W. Baskin), (Englewood Cliffs, New Jersey).

CLARK, C., [1951], Urban population densities; *Journal of the Royal Statistical Society, Series A*, 114, 490–496.

CLARK, P. J. and EVANS, F. C., [1954], Distance to nearest neighbour as a measure of spatial relationships in populations; *Ecology*, 35, 445–453.

CURRY, L., [1962], The geography of service centres within towns: the elements of an operational approach; *Lund Studies in Geography, Series B, Human Geography*, 24, 31–53.

DACEY, M., [1962], Analysis of central place and point patterns by a nearest neighbour method; *Lund Studies in Geography, Series B, Human Geography*, 24, 55–75.

DACEY, M., [1963], Order neighbour statistics for a class of random patterns in multidimensional space; *Annals of the Association of American Geographers*, 53, 505–515.

DIAMOND, D. R., [1962], The central business district of Glasgow; *Lund Studies in Geography, Series B, Human Geography*, 24, 525–534.

DUKE, R. L., [1964], Gaming-simulation and urban research; *Michigan State University, Institute for Community Development and Service*, (East Lansing, Michigan).

DUNN, E. S., [1954], *The location of agricultural production*, (Gainesville, Florida).

FAITHFULL, W. G., [1959], Ribbon development in Australia; *Traffic Quarterly*, 13, 34–54.

FORBES, J., [1964], Mapping accessibility; *Scottish Geographical Magazine*, 80, 12–21.

FOSTER, G. J. and NELSON, H. J., [1958], *Ventura boulevard: a string-type shopping street*, (Los Angeles).

GARNER, B. J., [1960], *Differential residential growth of incorporated municipalities in the Chicago suburban region*, (Mimeographed).

GARNER, B. J., [1966], The Internal Structure of Shopping Centres; *Northwestern University, Studies in Geography*, 12.

GARNER, B. J. and HARVEY, D., [1965], work in progress.

GARRISON, W., [1956], Allocation of Road and street costs; *Washington State Council for Highway Research, The Benefits of Rural Roads to Rural Property*, IV, (Seattle).

GARRISON, W., BERRY, B. J. L., MARBLE, D. F., NYSTUEN, J. D. and MORRILL, R. L., [1959], *Studies of highway development and geographic change*, (Seattle).

GETIS, A., [1963], The determination of the location of retail activities with the use of a map transformation; *Economic Geography*, 39, 1–22.

GODLUND, S., [1956], Bus service in Sweden; *Lund Studies in Geography, Series B, Human Geography*, 17.

GOTTMANN, J., [1961], *Megalopolis: the urbanised northeastern seaboard of the United States*, (New York).

GREEN, F. H. W., [1950], Urban hinterlands in England and Wales: an analysis of bus services; *Geographical Journal*, 116, 65–88.

GUNAWARDENA, K. A., [1964], Service centres in southern Ceylon; *University of Cambridge, Ph.D. Thesis*.

HAGGETT, P., [1965], *Locational analysis in human geography*, (London).

HARRIS, C. D., [1943], A functional classification of cities in the United States; *Geographical Review*, 33, 86–99.

HARRIS, C. D., [1954], The market as a factor in the localisation of industry in the United States; *Annals of the Association of American Geographers*, 44, 315–348.

HARRIS, C. D. and ULLMANN, E. L., [1945], The nature of cities; *Annals of the American Academy of Political and Social Science*, 242, 7–17.

HARTMAN, G. W., [1950], The central business district: a study in urban geography; *Economic Geography*, 26, 237–244.

HAWLEY, A. H., [1950], *Human Ecology*, (New York).

HAYES, C. R., [1957], Suburban residential land values along the C.B. & Q. railroad; *Land Economics*, 33, 177–181.

HORWOOD, E. and BOYCE, R., [1959], *Studies of the central business district and urban freeway development*, (Seattle).

HOYT, H., [1939A], *One hundred years of land values in Chicago*, (Chicago).

HOYT, H., [1939B], *The structure and growth of residential neighbourhoods in American cities*, (Washington).

HOYT, H., [1961], A method for measuring the value of imports into an urban community; *Land Economics*, 37, 150–161.

HOYT, H., [1964], Recent distortions of the classical models of urban structure; *Land Economics*, 40, 199–212.

HUFF, D. L., [1961], Ecological characteristics of consumer behaviour; *Regional Science Association, Papers and Proceedings*, 7, 19–28.

HURD, R. M., [1911], *Principles of city land values*, (New York).

ISARD, W., [1956], *Location and Space-economy*, (New York).

ISARD, W., [1960], *Methods of regional analysis: an introduction to Regional Science*, (New York).

ISARD, W. and KAVESH, R., [1954], Economic structural interrelations of metropolitan regions; *American Journal of Sociology*, 60, 152–162.

JONES, E., [1960], *A social geography of Belfast*, (London).

JONES, E., [1964], *Human geography*, (London).

KELLEY, E. J., [1956], *Shopping centres*, (Saugatuck, Connecticut).

KING, L. J., [1961], A multivariate analysis of the spacing of urban settlements in the United States; *Annals of the Association of American Geographers*, 51, 222–233.

KING, L. J., [1962], A quantitative expression of the pattern of urban settlements in selected areas of the United States; *Tijdschrift voor Economische en Sociale Geografie*, 53, 1–7.

KNOSS, D., [1962], *Distribution of land values in Topeka, Kansas*, (Lawrence, Kansas).

KOLB, J. H. and BRUNNER, E. DE S., [1946], *A study of human society*, (Boston).

LÖSCH, A., [1954], *The economics of location*, (New Haven).

MAYER, H., [1942], Patterns and recent trends of Chicago's outlying business centres; *Journal of Land and Public Utility Economics*, 18, 4–16.

MAYFIELD, R. C., [1962], Conformations of service and retail activities: an example in lower orders of an urban hierarchy in a lesser developed area; *Lund Studies in Geography, Series B, Human Geography*, 24, 77–90.

MORRILL, R. L., [1965], The negro ghetto: problems and alternatives; *Geographical Review*, 55, 339–361.

MORRISSETT, I., [1958], The economic structure of American cities; *Regional Science Association, Papers and Proceedings*, 4, 239–256.

MOSER, C. A. and SCOTT, W., [1961], *British Towns*, (London).

MURPHY, R. E. and VANCE, J. E. JR., and EPSTEIN, B. J., [1955], Internal structure of the central business district; *Economic Geography*, 31, 21–46.

MUTH, R. F., [1962], The spatial structure of the housing market. *Regional Science Association, Papers and Proceedings*, 7, 207–20.

NELSON, H. J., [1955], A service classification of American cities; *Economic Geography*, 31, 189–210.

OLSSON, G. and PERSSON, A., [1964], The spacing of central places in Sweden; *Regional Science Association, Papers and Proceedings*, 12, 87–93.

PHILBRICK, A. K., [1957], Principles of areal functional organisation in regional human geography; *Economic Geography*, 33, 299–336.

POWNALL, L. L., [1953], The functions of New Zealand towns; *Annals of the Association of American Geographers*, 43, 332–350.

PRED, A., [1963], Business thoroughfares as expressions of urban negro culture; *Economic Geography*, 39, 217–233.

RATCLIFF, R. U., [1949], *Urban land economics*, (New York).

REILLY, W. J., [1931], *The law of retail gravitation*, (New York).

ROTERUS, V. and CALEF, W., [1955], Notes on the basic- nonbasic employment ratio; *Economic Geography*, 31, 17–20.

SEYFRIED, W. R., [1963], The centrality of urban land values; *Land Economics*, 39, 275–285.

SIMMONS, J., [1964], The changing pattern of retail location; *University of Chicago, Department of Geography, Research Paper*, 92.

SIMON, H. A., [1955], On a class of skew distribution functions; *Biometrica*, 42, 425–440.

SKINNER, G. W., [1964], Marketing and social structure in rural China, Part I; *Journal of Asian Studies*, 24, 3–43.

SMAILES, A. E., [1947], The analysis and delimitation of urban fields; *Geography*, 32, 151–161.

SMAILES, A. E. and HARTLEY, G., [1961], Shopping centres in the Greater London area; *Institute of British Geographers, Transactions and Papers*, 29, 201–213.

SMITH, P. J., [1962], Calgary: a study in urban pattern; *Economic Geography*, 38, 315–329.

SMITH, R. H., [1964], *The geographical relevance of functional town classification*, (Mimeographed).

SMITH, R. H., [1965], Method and Purpose in Functional Town Classification; *Annals of the American Association of Geographers*, 55, 539–548.

STAFFORD, H. A. JR., [1963], The functional bases of small towns; *Economic Geography*, 39, 165–175.

STEIGENGA, W., [1955], A comparative analysis and a classification of Netherlands towns; *Tijdschrift voor Economische en Sociale Geografie*, 46, 106–112.

STEWART, C. T., [1958], The size and spacing of cities; *Geographical Review*, 48, 222–245.

STINE, J. H., [1962], Temporal Aspects of Tertiary production elements in Korea; *Urban Systems and Economic Development*, 68–88, (Eugene, Oregon).

THOMAS, E. N., [1960], Areal associations between population growth and selected factors in the Chicago urbanised area; *Economic Geography*, 36, 158–170.

THOMAS, E. N., [1961], Toward an expanded central place model; *Geographical Review*, 51, 400–411.

THOMAS, E. N., [1962], The stability of distance-population-size relationships for Iowa towns from 1900–1950; *Lund Studies in Geography, Series B, Human Geography*, 24, 13–29.

THORPE, D. and RHODES, T. C., [1966], The shopping centres of the Tyneside urban region and large scale grocery retailing; *Economic Geography*, 42, 52–73.

THÜNEN, J. H. VON, [1826], *Der Isolierte Staat in Beziehung auf Landwirtschaft und Nationalökonomie*, (Hamburg).

TOBLER, W. R., [1963], Geographic area and map projections; *Geographical Review*, 53, 59–78.

TUOMINEN, O., [1949], Das Einflussgebiet der Stadt Turku im System der Einflussgebiëte S.W. Finnlands; *Fennia*, 71, 114–121.

ULLMAN, E. L., [1941], A theory of location of cities; *American Journal of Sociology*, 46, 853–864.

ULLMAN, E. L. and DACEY, M. F., [1960], The minimum requirements approach to the urban economic base; *Lund Studies in Geography, Series B, Human Geography*, 24, 121–143.

VINING, R., [1955], A description of certain spatial aspects of an economic system; *Economic Development and Cultural Change*, 3, 147–195.

VOORHEES, A., [1955], A general theory of traffic movement; *Proceedings of the Institute of Traffic Engineers*, 46–56.

WATSON, J. W., [1955], Geography: a discipline in distance; *Scottish Geographical Magazine*, 71, 1–13.

WINGO, L. JR., [1961], *Transportation and urban land*, (Washington D.C.).

WINSBOROUGH, H. H., [1962], City growth and city structure; *Journal of Regional Science*, 4, 35–49.

YEATES, M., [1965], Some factors affecting the spatial distribution of Chicago land values, 1910–60; *Economic Geography*, 41, 57–70.

ZIPF, G. K., [1949], *Human behaviour and the principle of least effort*, (Cambridge, Massachusetts).

CHAPTER TEN

Models of Industrial Location

F. E. IAN HAMILTON

The major contributions to location theory have been made by economists, mostly with a view to integrating location studies with the main body of economic theory. The very word 'location', however, implies the existence of spatial relations, inter-relationships, and patterns, so that models of industrial location are, by definition, part of geography. Of course, industrial production *is* an economic activity. Yet unless it is so integrated that the procurement of materials, the manufacturing process, and the marketing of the product all take place at the same point (as in an intermediate stage of a factory process), the spatial separation of materials and markets involves transport to overcome distance and so effect the geographical linkage of supply with demand. Geographers have published hundreds of works giving descriptions and analyses of empirical evidence concerning the reasons for, the effects of, and changes in, the location of industries. In general this evidence is unbalanced in favour of a few branches of industry, a few world regions or nations, and an idiographic approach which tends to dissipate, rather than to integrate, the body of location theory and practice. It is not difficult to see why this should be so.

The term 'industry' describes a wide range of activities. These are as diverse as the quarrying of chalk, the smelting and refining of metals, the assembly of electronic equipment, and the ubiquitous supermarket. Industries can be classified into four groups, therefore, according to the operational process that they use: respectively, extraction, processing, assembly, and service. Each group requires specific inputs (materials, labour, capital) from specific sources for its operation and provides specific outputs for purchase in specific markets and market areas. The locational requirements of industries in each of the four groups differ as a result; and within each group these requirements are as a 'theme with variations'. Beyond this level, the number of location variables increases rather than decreases. Industries are not only heterogeneous activities; each branch is susceptible to diverse means of organization. The relative importance of the location factors involved, and hence the final location chosen, is likely to vary, even for the same branch of industry, as between different scales of production and between the one-plant firm, the intra-industry multi-plant corporation, the multi-industry firm with one or several plants and the state-managed enterprise. Further interpretations of the reality of location arise from analyses that account changes in time and in technique, and that compare industrial development in the light of variations in the size of regions and nations and in their resource endowment, population and markets.

This essay does not purport to fill in gaps in empirical research. Rather it presents some general and particular conceptual models of industrial location phenomena, using and criticizing both the familiar and the unfamiliar, and suggests models of location in the context of capitalist and socialist economic systems, settlement hierarchies, history, regions, nations and continents.

THE CHANGING CHARACTER OF MODEL TECHNIQUES

We begin, however, by considering certain progressive changes which have occurred during the last century in the ways in which models have attempted to reach a solution to the location problem.

The models that pre-date the First World War were generally simple constructions in which few variables were assumed and rigid mathematical proof (usually involving geometry) was applied to determine some optimum location. Launhardt (1882), for example, developed the location triangle for a given factory in which two points represented the sources of the materials it used and the third point was the market where it sold its products. Launhardt determined the optimum location as the point within the triangle at which the shortest lines (representing distance) met from the three points. Later (1885) he presented a more complex solution, the 'pole principle' (see Palander, 1935, p. 143). In the same work Launhardt demonstrated that ideal market areas could be hexagonal (Lösch, 1954, p. 114), but he proved that in practice they tended to be irregular polygons with boundaries as curves of the fourth degree. In the twentieth century, Weber (who was deeply impressed by Launhardt's work) and Palander have made extensive use of geometry.

Alfred Weber (1909) made the first important break with geometry, however, when he introduced the 'isodapané', or a line joining points of equal cost. This technique permitted greater diagrammatic flexibility and greater approximation to reality. The retreat from geometric determinism had begun. Palander (1935) and Hoover (1937) used isodapanes (and other isolines) extensively to show the locational irregularities in procurement and distribution costs that resulted from the varied pattern of different transport media with different cost/distance ratios. These and other examples were often translated into graphs which were popular among economists.

Like geometry, however, the iso-line and graph techniques suffered severe limitations. They remained satisfactory only so long as the few variables that *were* considered were reduced to fixed points, and distance was expressed by the length of a line, by the radius of a circle, or by some iso-interval. Hardware models never seemed to be popular. Only those of Launhardt (a mechanical weight model) and Weber (similar, but developed on a Varignon frame) had any significance and they were primitive, though no doubt adaptable to more variables with effort.

The need for a new method to cope with several location variables became apparent when it was realized that neither were supply or demand point-formed nor were industries organized only in one-plant firms. Hitchcock responded to this need by devising (1941) an algebraic formula for minimizing transport in the distribution of a product from several manufacturing plants (located at different points) to a large number of market areas. This was the origin of the formula:

$$y = \Sigma a_{ij} x_{ij} \text{ summed for } i = 1, 2 \ldots m,$$
$$\text{and } j = 1, 2 \ldots n,$$

where y is the total cost, a_{ij} is the cost of transporting one ton product from the ith factory to the jth market, and x_{ij} is the number of tons to be moved. This linear programme technique is eminently applicable to industrial location problems if the optimum (maximum or minimum) solution can be expressed algebraically as a linear equation. Kuhn and Kuenne (1962, p. 21) assert that 'it is not fortuitous that the origins and development of programming techniques are so closely tied to the problem of spatial economics, and that these techniques have resulted in the advance of breakthrough proportions in obtaining solutions to formerly insoluble spatial problems'. However the application of the technique is not without its problems. Koopmans and Beckman (1957) developed a model to assign n factories to n locations in such a way as to maximize the combined profits of the plants. They found that as the number of variables considered increases, the programming method becomes too cumbersome for satisfactory solution of the 'assignment' or 'allocation' problem. Garrison (1959, p. 479), for instance, shows that 120 possible location patterns must be compared in assigning only five plants to five locations! Similarly, Bos (1965, pp. 76–84) found that some 10^9 linear programmes were required to achieve the optimum spatial dispersion of three vertically-integrated industries over ten possible locations. He concludes (Bos, 1965, p. 84) that further progress in this field requires the development of 'an operationally efficient (algorithmic) method of solution for mixed integer programmes.' An algorithm has been applied to a Weberian location problem involving $n = 4$ by Kuhn and Kuenne (1962); they assert first, that it is superior to all geometrical, iso-line, and physical analogue methods and second, that even with more than four variables the calculations involved 'are well within the capability of the desk calculator for large numbers of elements' (1962, p. 21).

For reasons that are set out in the next section, however, models are becoming more elastic in their approach to location problems. Greater realism can be achieved if these methods are complemented by stochastic models which in part are based on game theory, 'random walks', Monte Carlo simulation, and Markov-chain analysis.

IN SEARCH OF REALITY

An attempt is made in Figure 10.1 to construct a model which forms the cornerstone of industrial location model-building. It is logical to start with the three kinds of entrepreneur that decide location in the real world: the private capitalist, the corporate capitalist, and the State administration (local, republic, national, or federal). The model thus clearly distinguishes three different approaches to the location problem. These vary in importance both in time and place. Private capitalists determined the location pattern during the industrial revolution or during the early stages of industrialization while capital requirements were small. Joint stock companies and corporations have become more important entrepreneurs in the capitalist sector in the twentieth century. And now the State not only determines location choice in the communist world, but it plays an important part also in location decisions in the mixed economies of the non-communist world. These entrepreneurial differences indicate that the location models of Launhardt, Weber, and Hoover need not contradict those of Lösch or the socialist school, and vice versa; rather they may be complementary.

Profit motivates decisions in the capitalist sector, while such less easily quantified motives as 'the national interest', 'social cost' or 'social benefit' colour State location choices. Our model specifically omits the word 'maximum' from either criteria since this expresses an ideal not a reality. Greenhut and Colberg (1962, pp. 43–44) argue convincingly that, given exactly the same business or plant to locate, no two private entrepreneurs would judge alternative locations by the same quantitative 'maximum' profit. One will seek a large profit; another will be content with some profit. Indeed, no two states judge social cost-benefit in exactly the same way. Thus divergence of entrepreneurial opinion allows greater flexibility of location choice by increasing the range of 'optima'. Indeed, 'most human decision-making, whether individual or organizational, is concerned with the discovery and selection of *satisfactory* alternatives; only in exceptional cases is it concerned with . . . optimal alternatives' (March and Simon, 1958, pp. 140–141). The one, however, is not necessarily inferior to the other (Odhnoff, 1965). Circumstances that are peculiar to each kind of entrepreneur also influence their

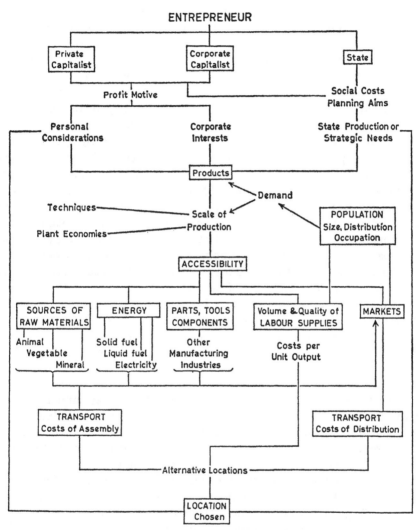

10.1 A basic model of the factors influencing industrial location decisions.

interpretations of the motive and hence their final choice of location. Personal considerations of either a partial economic nature (good financial contacts) or a purely social character (climate, hobby possibilities) are more likely to influence the private entrepreneur in choosing location than the corporation or State. More important may be his development of some small or growth industry in chance or random locations in response to a combination of favourable local conditions (inventions, laws, capital), a willingness to operate in the known environment of the 'home town', and an unwillingness to try an unknown locality elsewhere (Katona and Morgan, 1952). Evidence suggests that few private capitalists undertake comparative cost analyses to justify or refute their 'hunches' (McLaughlin and Robock, 1949, Chaps. 2–3; Alkjaer, 1953, p. 90; Klemme, 1959, pp. 71–77; Estall and Buchanan, 1961, p. 18; Goss, 1962, p. 16; Luttrell, 1962). When a firm grows into a corporation, personal factors are subordinated to the interests of shareholders and to the efficiency of plant in a highly competitive market: location decisions are influenced by corporate interests which include the efficient spatial, technical and economic relationships of new plant to existing plants, and the division of the market to ensure adequate supplies of products to existing or potential market areas. Large firms endeavour to compare location costs because of the greater risks they run on account of their scale.

As our model shows, the State can bring its social and planning policies to bear upon decisions in the capitalist sector of a mixed economy; these may take the form of negative and positive 'push-pull' forms of control (Manners, 1965, p. 153). Theoretically, the State ought to be absolved of personal influence in location policy; in practice powerful individuals or groups may steer important projects to particular areas even where planning is strictly centralized. Generally, however, the considerations that influence State interpretation of 'social benefit' or 'national interest' may include: the development of material-oriented industries to reduce the excessive export of mineral resources in raw or semi-processed form, or of manufacturing to reduce the dependence upon imports of capital-goods (Hamilton, 1962); the strategic allocation of projects both to many and to isolated areas (Hamilton, 1964A, pp. 59–61); and the location of industries to develop backward, under-privileged, or problem regions (Hamilton, 1964B, pp. 78–80).

Extraneous elements are introduced into the next stage of the model: demand (influenced by population size and occupation structure), techniques, and plant economies. These factors do not cause 'noise' because together they condition the line or range of production (and indirectly, therefore, the 'input mix'), they establish how many plants of optimum scale are required to serve the market, and they indicate how far the production process is technically divisible or indivisible and thus amenable to spatial dispersal or concentration. The scale of operation chosen is an important consideration in examining the accessibility that locations offer to supplies of materials and labour, and to markets. The importance of the range of products is now revealed: in part conditioning scale, in part conditioning cost structure where complex industries, e.g. chemicals, are concerned (Isard *et al.*, 1959). Entrepreneurs should then compare the summed totals of production and transport (procurement) costs of raw materials, fuel, power and components at, the costs of labour inputs in, and the costs of distributing the products to the market from alternative locations (Fig. 10.1). The model is comprehensive, therefore; yet *per se* it cannot indicate the relative importance of each location factor either for industries in agregate or for an industry in particular. Nevertheless, it does hint at a variety of input and output characteristics, and hence at various location relationships. The segregation of animal, vegetable and mineral raw materials suggests differing methods and sources of procurement and different degrees of 'perishability'; and the distinction between 'raw materials' and 'parts, tools and components' points up fundamental dissimilarities between primary (more nature-tied) and secondary (man-made and more 'footloose') sources of materials. The variable transportability of energy, and different labour characteristics, are also indicated. Finally, the model links the sources of materials, energy and components with 'markets' to emphasize the diversity of markets and market areas for capital goods, and it links population with 'markets' to underline the existence of distinct consumer markets. These details are essential in view of the frequently imprecise usage of the terms 'materials' and 'markets' in location literature. Their distinction, moreover, adds effectively to the beauty of the model.

As Figure 10.1 indicates, the final selection of location is made according to a comparison of economic costs, but the decision may be swayed by the individual's personal desires or by some particular State planning requirement. National or regional economic development and planning policies affect the location decision-making model in one further respect. The selection of a certain regional allocation of plants to fulfil planning needs may well have a feedback effect upon the scale and range of outputs of the plants concerned and hence their precise location. In other words, the increasingly important macro-economic approach to location choice demands more frequent reappraisal of the micro-economic approach. Once a locality is selected, a site must be found for the plant. The site that is chosen must usually ensure the availability of: adequate land for plant, stores, access facilities, and future expansion; adequate water supplies for processes or products; the right kind and quality of transport service; sufficient labour within an easily accessible area; and the preservation of amenity as stipulated by government regulations regarding site relationships with other land uses (agricultural, forestry, extractive, residential, recreational), atmospheric pollution, noise, danger or strategy.

It cannot be denied that the economics of procurement, and production distribution combined is *the* important – even deciding – location considera-tion. To infer that it is the only one, however, and accordingly to explain loca-tion mechanically, as many location models do, is to deny that man is human. The location patterns that emerged in the period between the industrial revolution and the application of comprehensive planning (whenever that may be) reveal that 'accident, habit, . . . and man's impulse to conquer his environment and to canalize his random impulses into orderly activities produced . . . nothing less than the empire of muddle' (Mumford, 1934, pp. 194–195). When knowledge was restricted and individualism was strong, man's gregariousness led to specialization which, under the conditions of an expanding economy, produced specialized agglomerations so that 'a variety of regional opportunities were neglected, and the amount of wasteful cross-haulage in commodities that could be produced with equal efficiency in any locality was increased . . .' (Mumford, 1934, p. 171). Moreover, man's think-ing on the location of industries is formed under pressure from a vast array of human practices, prejudices, habits, laws and systems. Nationalism, im-perialism and other political realities provide the true framework within which the location, distribution and spread of industries are encouraged, facilitated, restricted or prohibited (Odell, 1963; Hamilton, 1964A, pp. 46–64). Similarly, the attitude of communities, as Wallace and Ruttan (1960, p. 140), Paterson (1963, p. 130) and Hiner (1965, pp. 23–24) point out, can attract or repel or select industries on both the regional and local levels. 'Anomaly' industries or locations may result from specific ethnic distribu-tions (Alexander, 1963, pp. 308–309). The early start of an industry in either a rational or a random location with some initial advantage (Pred, 1965), given dynamic entrepreneurship, may set in train the development of auxiliary industries, services and urban infrastructure that, at a later stage, ensures the inertia of the industry. If entry into the market is difficult for competitors, more advantageous locations may remain unexploited. Perhaps, however, the possibilities of 'sub-optimal' locations are nowhere as great as for industries manufacturing for the ultimate consumer. Here, product dif-ferentiation, as Greenhut (1964, Chap. 11) shows, distorts the shape and the extent of market areas and creates 'product islands'. Equally, advertising (Alexander, 1963, p. 310) may give extra marketing advantages (and profits) to a firm or a locality which, in cost terms, is less favourably located than rivals are. Even if the economic geographer is convinced that transport costs explain location patterns, there is no direct relationship between transport rates and geographic space (Alexander, 1963, p. 473). Special rate-fixing and pricing policies may often distort location patterns in practice.

While, therefore, the beauty of simple economic interpretation is the attraction of a model's vital statistics, the reality of her heart reminds one of the dangers of ignoring aspects which cannot be expressed in figures!

EARLY MODELS OF INDUSTRIAL LOCATION

It is against this background that one must examine early attempts at model-building. Many of the ideas at the core of location theory have their roots in the nineteenth century. To the early British economists – Adam Smith, J. S. Mill and David Ricardo – the location of industry was partly a function of the spatial distribution of agricultural surpluses for feeding industrial labourers and for processing. Writing between 1775 and 1810, Smith develops the threads of a model for industry at a time when technical innovations and canal transport were changing the face of Britain. Around mid-century, Karl Marx (1850) revealed what might be termed a 'capital localization-location' model which was framed in a world of international political inequalities. The last quarter of the nineteenth century saw a growing interest, especially among theorists in Germany, in location patterns within regions. Schäffle (1878) suggested the outline of a gravity model. According to this, industries develop chiefly in or near larger towns which, as markets, attract industry in direct proportion to the square of the distances between them:

$$M_{ij} = P_i P_j (d_{ij})^{-2}$$

for two centres i and j where M_{ij} is the market attraction of the two towns, P_i and P_j represent their populations, and d_{ij} is the distance separating them. Any significant deviation from this pattern Schäffle attributes to the location of industries near raw material or fuel resources. The model seems to explain broadly the patterns of 'ubiquitous' and of market-oriented sporadic industries; it would not necessarily function well, however, in explaining much sporadic manufacturing. In a wider context, the model implies that there is a greater localization of industry the larger is the urban population, the shorter is the distance to the market, and the closer are large cities to one another. It provides theoretical support for the more recent empirical findings of Bogue (1949) and Duncan (1959) in the United States.

Perhaps the most influential models of the nineteenth century were those developed by Launhardt.[1] He argued (Launhardt, 1882) that differences in the costs of production and prices between production centres and in the rates of transporting products modified the size and shape of the sales areas that could be supplied from those centres. Later Launhardt (1885) extended this idea to show the crucial importance of ton-mileage (weight × distance) × the transport rates at alternative locations in determining the costs of production. He concluded that entrepreneurs would develop industries in the locations with the *least costs*. It is significant, however, that Launhardt, like Schäffle before and Loria after him, never lost sight of the importance of the

---

[1] The work of Launhardt had a profound influence upon Weber's location theory; many of his ideas have been taken up by Palander, Lösch, and Isard more recently.

market. Thus Achille Loria (1888 and 1898) asserts that industries are located with reference to the market areas that they serve, with the exception of those industries which use bulky materials losing most of their weight in processing. The less these raw material-oriented industries needed to transport bulky materials, he argued, the larger the market area their products could command at a given price. Although Loria makes little reference to pricing policies, his model works on the premiss that entrepreneurs seek the locations that will give them *maximum profits*. This gifted Italian economist was also among the first to recognize that labour was an important location factor. He assigned industries with high labour inputs in the production process to predominantly agricultural areas. Röscher (1899) made a related observation, explaining that labour not only attracts industries to areas of high population density and localized skills, but also encourages industrial development in areas with 'workers . . . (in need of) supplementary income' (Krzyżanowski, 1927, p. 279).

THE WEBER MODEL: MERITS AND DEMERITS

Although the idea of the least-cost location model originated in the work of Launhardt in the nineteenth century, the energy of Alfred Weber – in extending, modifying, and propagating it – made it all-persuasive in the twentieth century. Weber (1909) frames his model in an isolated state where natural resources for processing conform to a Thünen-ring system around a number of given market centres. These assumptions imply the existence of spatial variations in economic advantage with respect to both supply and demand. Natural resources comprise 'sporadic' materials (e.g. mineral fuels and metallic ores) and 'ubiquitous' materials (e.g. water, sands, clays, stone) which have either 'gross' weight if they lose weight in processing, or 'pure' weight if they lose none of their weight. In this way he distinguishes between more localized and more dispersed materials and between 'less mobile' and 'more mobile' materials.

Weber stresses that, within this heterogeneous environment, entrepreneurs will locate industries at the points of least cost in response to three general location factors – transport and labour (as inter-regional factors), and agglomeration or deglomeration (as intra-regional factors). He assumes that transport costs are a function of weight and distance (ton-miles). The point of least transport cost, therefore, is that at which the combined weight movements involved in assembling materials from their sources and in distributing products to their markets is at a minimum. Weber (1929, pp. 48–75) devised a 'material index', given as the weight of localized material inputs divided by the weight of the product, to indicate whether the point of 'movement minimization' (Haggett, 1965, pp. 142–148) would be near the source of materials (an index >1) or near the market (an index <1). If an industry has a high labour coefficient (according to Weber, the ratio of labour cost to the combined weights of inputs and outputs), Weber argues that labour will attract the industry to a location other than the movement minimization one if the savings in labour cost per unit output exceed the extra transport costs involved. Substantial agglomeration or deglomeration economies could encourage entrepreneurs to 'deviate' from the minimum transport and labour cost locations and to develop their industries at a third 'minimum' point. To Weber, agglomeration economies arose from local internal or external (linkage) economies of scale, labour skills, bulk buying and selling to minimize stocks, and infrastructural benefits. He stressed that, by contrast, rising land prices in growing urban areas encourage deglomeration.

Weber's model distinguishes at least fourteen hypothetical types of industry (1929, pp. 61–66) regarding weight movement, labour coefficients, and agglomeration economies; these and the 'orientations' Weber suggests are given in Table 10.1.

TABLE 10.I

Inputs, Location factors, and Least-Cost Locations for Weber's Hypothetical Industries

Industry	Ubiquitous Materials	Inputs Sporadic 'Pure' Materials	Inputs Sporadic 'Gross' Materials	Labour	Agglomeration	Deglomeration	Materials	Intermediate	Market	Labour	Agglomeration	Deglomeration
				(Other Factors)			(Orientation)					
A	1								*			
B	2+								*			
C		1					?	?	?			
D	1+	1							*			
E		1					z					
F		2							*			
G			2=					*				
H			2≠ +				*					
I	1+		1				?	?	?			
J	2+											
K	unspecified			*						*		
L	"	"	"		*						*	
M	"	"	"	*							*	
N	"	"	"			*						*

The symbols used are:

+ for 'or more' materials
= for equal weight losses for gross materials
≠ for unequal weight losses for gross materials
* for definite locations
? for alternative (equally viable) locations

This clearly reveals the main features of Weber's model. Industries A, B, and D, that process ubiquitous materials, are oriented to the market since any other location would involve the unnecessary costs of transporting ubiquitous materials in the form of a bulkier product. Weber argues that industry F, using two localized 'pure' materials, will locate at the market for the same reason. Where differentially weight-losing materials form the only inputs then entrepreneurs will locate their industries near the sources of materials, ideally, that lose most weight (industries E and H). Where two or several gross materials lose equal weight (industry G) entrepreneurs will locate at some intermediate assembly point. Weber solved this location problem by using Launhardt's location triangle, while Pick (Weber, 1929, p. 229) and Cotterill (1950, p. 67) demonstrated it respectively by the Varignon frame and the force table analogue. The location optima, however, are not so obvious either for industry C which manufactures a 'pure' sporadic material incurring (à la Weber) equal transport costs wherever the industry is located, or for industries I and J which process both localized weight-losing and ubiquitous materials. In this last case the 'minimum movement' location will depend upon the input ratio of gross materials (and their weight-losses) to ubiquitous materials. Industry K, with a high labour coefficient, is located in the cheapest labour location, industries L and M are linked and are attracted to a common centre on account of agglomeration economies, and industry N seeks out an 'out-of-town' location because of large land requirements. Clearly, the dominance of market-oriented locations in Weber's scheme refutes those who criticize him for a preoccupation with material-orientation.

By disproving the reality of Weber's basic *assumptions*, however, critics have undermined the very essence of his model. His disregard for the imponderables of human beings set out earlier is an example; this is supported in a devastating critique by Hamill based on empirical evidence from the Oregon timber industry (in Hamilton, 1964C, p. 234). To assume, for instance, that raw material sources and markets are fixed points is to ignore the geographical conditions of supply to agricultural and forest industries, and of demand generally. It also takes no account of possible spatial changes in supply or demand. Weber presupposes an unrealistic framework of perfect competition and given market conditions. This led him to analyse only cost factors and to maintain that, by definition, the least-cost location was also the maximum profit location. Production is also assumed to be given so that 'production cost has no importance and transport cost is the only factor for determining the optimum flow' (Ghosh, 1965, p. 44). Moses (1958) proves conclusively that optimum locations depend upon optimum production levels. Weber erred also in trying to reduce real differences in raw material production costs at source to a transport problem by inventing fictitious distances; this only confuses the problem. Subsequent location theorists (Hoover, 1937,

p. 52; Isard, 1956, p. 108; Fulton and Hoch, 1959, p. 52) have stressed the fallacy of the assumption that transport costs are proportional to distance. Thus Isard (1956, pp. 105–108) and Alexander (1963, pp. 473–475) show that freight rates tend to be convex and stepped with distance partly because they include fixed terminal and loading charges (Hoover, 1937) which are unaccounted for by distance. These authors emphasize, moreover, that, weight for weight, raw material freight charges are usually lower (sometimes much lower) than those for products. Yet Weber states (Friedrich, 1929, pp. 42–48) that transport costs – in addition to weight and distance – are determined by the type of carrier, the size of shipment, the nature of the region, the density of the transport network and by the nature of the goods to be carried (their quality, value, and perishability).

Dean (1938, p. 19) asserts that Weber '. . . seriously overestimated the determinate influence upon location of weight-losing materials when they were not dominant, and underestimated the attractiveness of pure materials which are never dominant.' This throws into sharp focus the lack of precise industrial input data in the model. In any case, industries A, C, and E are fictitious in the sense that no industry uses only one material input; even most simple processing activities require at least two. Smith (1955) has criticized Weber for including the use of coal in the weight index because it confuses raw-material with fuel orientation; when coal is excluded, however, the material index bears little relation to real location orientations, particularly when the weight lost is less than 75 per cent. These conclusions seem to support Dean's observations regarding the variety of influences of 'gross' and 'pure' materials on location choice.

Unreal assumptions and important inconsistencies weaken the force of Weber's labour and agglomeration analyses. Hoover (1937, p. 63) suggests that the 'labour coefficient' ought to express a relationship between labour cost and other costs (not weights) to gauge its importance. In his assumptions of unlimited labour supply and fixed regional labour cost differentials, Weber overlooks the facts of labour mobility and of labour shortage in growth areas. Sombart (1919) sees the recognition of wage differentials – the essence of capitalism – as a negation of Weber's claim to universality. Critics lay bare his misconceptions on agglomeration, pointing to the naïvity of geometric analysis and to the curious conclusion that agglomeration reinforces the attraction of cheap labour locations (Friedrich, 1929, pp. 161–162). Had Weber adopted more meaningful assumptions he might have concluded that agglomeration arises from the intensive exploitation of large and localized natural resources and major transport nodes, from urban economies, and from the growth of large city markets (Isard, 1956, p. 187). Yet labour *skill* may encourage the agglomeration of particular industries or groups of industry.

The Weber least-cost location model is, then, very 'noisy' in terms of its abstraction from real conditions. As it stands, its utility lies in its conceptual distinction of the relative importance of 'ubiquitous' and 'localized' materials, the inputs and outputs to be moved, transport and non-transport (labour, agglomeration) factors and the possible different 'orientations' of industries to materials, labour, and markets.

WEBER'S MODEL REFORMED: AN APPROACH TO REALITY

These criticisms lay the foundation for improving Weber's model by modification and extension. It appears that all *raw* materials are 'gross', although crude oil loses very little weight (Alexander, 1963, p. 345); 'pure' materials are those semi-finished industrial products (i.e. refined metals, plastics, metal components) that become the materials of fabricating, finishing, and assembly industries. In the reformed model Weber's ton-mileage transport criterion is replaced by one which compares value with weight, weight change in processing, bulk, perishability, fragility, and haulage rates. This still confirms Weber's thesis concerning ubiquitous materials since these 'abundant' materials command low – often very low – values in relation to their weight or bulk; they are thus expensive to transport. Nevertheless there may be contrasts in the types of raw material used and in the methods of mining them. These differences influence production costs directly through material procurement costs, and indirectly because they affect plant size and production processes; this gives a market advantage to low-cost producers as, for example, to manufacturers of Fletton bricks in England (Gleave, 1965). Localized materials and fuels are scarcer and command higher values, but these values vary widely according to the physical properties of, and man's uses for, such materials (e.g. iron compared with gold).

Industries, then, will tend to be located nearer the sources of their materials and fuel (or dominant input) the greater the proportion of unusable waste, the greater the bulk (perishability or fragility), the higher the freight charges, and the lower the value – weight for weight – of the raw materials compared with the products. The lower the weight loss, bulk, and the haulage rates, and the higher the value of the raw materials used, the more likely market-orientation will be. A number of location possibilities exist for industries that use 'pure' materials which neither lose nor gain weight in manufacture. Entrepreneurs will locate such industries nearer the source of the dominant material input if this has a low value/weight (bulk) ratio. Materials with high value/weight ratios are more transportable so that labour, market, or intermediate locations are more favoured for industries using them (Ross, 1896, p. 258). Industries tend to be market-oriented if they process divisible materials (oil), or assemble several 'pure' materials giving a substantial gain in weight or bulk (cf. value) in the product. Finally, the greater the gain in weight, bulk, perishability, or fragility, the higher the costs of transport, and the lower the value of the product, the greater is the tendency for entrepreneurs to locate an industry as near the centre of the market as possible to minimise distribution costs.

These principles may be modified or extended by factors which Weber did not take into account. For instance, entrepreneurs will locate industries near the cheapest adequate power-supply source if the ratio of power costs to other costs is very high. Those managing industries that depend upon labour skill, fashion, speedy contact or personal service (Smith, 1952; Martin, 1964) will choose locations where these give the greatest apparent advantages. Similarly, transport rate structures and facilities may distort the emerging patterns. Higher rates for products than for materials tip the balance further in favour of market-orientation; so does the availability of cheap (water) transport for carrying bulky low-value materials or of pipeline transport for other materials. Moreover, transport rates (Hoover, 1948) tend to 'polarize' industries at terminals, at the sources of materials or at the centres of markets, and to discourage intermediate locations unless favourable 'in-transit' rates are offered. The insistence upon cost that characterizes this 'reformed' model establishes a significant shift of emphasis from Weber's original concept of 'nearest material source' or 'nearest market' to the 'cheapest material source' and the 'most lucrative market' (Beckmann and Marschak, 1955, p. 333). This last idea provides a link with the final section, that concerns the market.

Weber laid the foundations for extending his own model, with regard to the market, in Chapter VII of *Über den Standort der Industrien* (1909). He suggests that an economy comprises a hierarchy of four 'strata' – agriculture and forestry, mining and primary processing, manufacturing, and tertiary services – that are interrelated as sources of supply and demand. This contributes also to an extension of the agglomeration concept to include industrial symbiosis and agglomeration at markets or sources of raw materials. The geographer can interpret each stratum as one market area or as a series of market areas either for industry as a whole or for particular industries. When combined, the 'strata' would give an irregular pattern of highly concentrated markets – where resources (e.g. coalfields) or nodes were localized – over a continuous and weaker market surface which reflected more extensive land uses and more dispersed settlement and population. The coalescence and overlapping of market areas with each other and with the centres or areas of materials-supply would result, as in reality. Industries, therefore, will procure their inputs from, and distribute their outputs to, a point, many points, or, most usually, an area or many areas which tend to adjoin or overlap.

The location problem here seems to resolve itself into a comparison of the importance of linear (point to point) and areal (point to area and vice versa) transport costs of procuring the materials and distributing the products (Lindbergh, 1953, p. 30). To obtain areal transport costs, Bos (1965, p. 34) shows that 'for practical purposes . . . a not too irregularly-shaped area can be approximated by calculations of cost for a circular market area of the same size'. Stefaniak (1963, p. 433) proves that a central location in a circular area minimizes distribution costs; these rise steeply if off-centre locations are chosen. Where transport cost differentials at alternative locations *are* decisive, therefore, an industry will be located at that centre where the combined areal transport costs of materials *and* products plus the linear haulage cost of transporting *either* materials *or* products from the supply area centre to the market area centre are at a minimum.[1] Naturally, an areal market (material-supply) strengthens the tendencies toward market-orientation (material-orientation), according to the principles established earlier, given point-formed material supply sources (markets). Yet the market area will always tend to have the greater attraction even where industries procure materials from an area; it will also 'pull' an industry somewhat away from materials if that industry is material-oriented. There are two reasons for this. First, areal as opposed to linear distribution magnifies the difference between the higher freight rates for products and the lower rates for materials of the same weight and bulk; the same effect results from the diseconomies of short-haul distri-bution in an area (especially where road transport is limited) compared with the economies of long-haul linear transport. Second, the *size* of demand in the market area is of crucial importance in determining distribution costs (Lind-bergh, 1953, p. 30); but since the 'maximum profit' or 'maximum benefit' motive is more realistic than the 'least-cost' motive, industries will locate so as to tap the largest sales potential within the market area (Greenhut, 1960; Greenhut and Colberg, 1962, pp. 29–42), or to maximize some social benefit. This location may give them a large profit, or yield larger benefits, even though the costs of assembly and distribution are higher than at the least cost location. It should be possible to graft the 'market-potential' concept (Harris, 1954; Dunn, 1956; Isard *et al.*, 1960) on to this location model. In the last analysis, Weber's model can be improved if his assumption of given markets and population distribution is replaced by a dynamic approach which recognizes that relatively large-scale industrial development itself changes the geography of market potential. The location of a large, or several smaller, new plants in or near existing industrial areas speeds and strengthens the agglomerating forces and weakens the ability of unindustrialized areas to attract industry. If planning or some less direct device encourages deglomeration through the location of new plants in areas which lack industry, the process may be re-versed for in time the growth of employment and industry creates a more attractive local market for consumer and capital goods in such areas, provid-ing a far more satisfactory basis for counteracting agglomeration. Such a consideration provides a link with models of location under capitalism.

[1] Or at the point where costs *are thought to be* at a minimum.

INDUSTRIAL LOCATION UNDER CAPITALISM

Locational interdependence

Neglect of the market influence by the Weber school stimulated economists, especially in America, to consider location under the real conditions of oligopoly, duopoly, and monopoly. The entrepreneur's motive for location under these conditions is to control the largest segment (or the whole) of the market area at prices that yield him the greatest profit. The duopolist or oligopolist, therefore, must pay special attention to the location and markets of his rivals. Each firm's location is 'interdependent' (Greenhut, 1952, pp. 526–538) with those of other firms in the industry. Hotelling (1929) concludes that two firms locate their plants near the centre of the market if demand is inelastic; they will locate further apart if it is elastic. Lösch (1940), Greenhut (1957), and Devletoglou (1965) disagree; they contend that duopolists locate plants to divide, yet to gain more than half of, the market area (in sales volume); to do so they must be located at the quartiles along a line

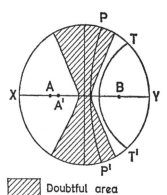

10.2 The locations chosen by duopolists in a circular market area are at A and B. Fetter's theory is introduced. If all costs are equal to both manufacturers, they will divide the market equally. Lower *production* costs at A, however, would enable the producer located there to extend his market to the line P – P'. Lower *transport* costs from A would permit expansion to the line T –T'. Theoretically, a location at A' has the same result.

Doubtful area

through the market (locations *A* and *B* in Fig. 10.2). Yet wherever monopolistic competition occurs there are always areas (shaded in Fig. 10.2) where the market is doubtful (Greenhut, 1957, p. 86; Devletoglou, 1965, pp. 144–149), i.e. where competitors vie for the market. Even so, each manufacturer may retain a monopoly control of the areas that lie farthest away from rivals i.e. area *A–X* for *A* and *B–Y* for *B* in Figure 10.2. An entrepreneur may become a spatial monopolist if his plant (and the market it serves) is separated geographically (and economically by high transport costs) from the market areas of other firms in the industry e.g. the Pacific coast of North America (Harris, 1954). Smithies (1941) shows that the monopolist will either centralize production in relation to the market (in industries where savings in material movement or agglomeration economies far exceed the costs of distributing products); or he will divide it into equal market areas served by separate plants (where 'ubiquitous' or 'pure' materials are used or where the product bears high haulage charges on account of low value, weight, bulk, perishability, or fragility). If production is centralized, the plant may not be located at the mathematical centre (X in Fig. 10.3); it will be in a median location (A in Fig. 10.3) with access to the greatest number (or purchasing power) or buyers (Quinn, 1943). This model can be grafted on to the reformed Weber model, especially with regard to market-oriented industries.

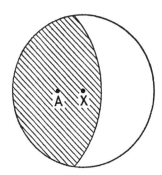

10.3 The median location. All buyers are located in the shaded area of the region concerned.

The maximum profit model

The Weber school eclipsed Loria's maximum profit concepts until Lösch published *Die räumliche Ordnung der Wirtschaft* (1940). He asserts (1954, pp. 28–29) that 'seeking the place of least cost . . . is as absurd as to consider the point of largest sales as the proper location'. Lösch attempts to find the maximum profit location by comparing the costs of production at, and the market area which can be controlled from, alternative locations. Given the framework of oligopolistic competition, the location chosen may well not be the least-cost one since profits depend more upon sales revenue within an area than upon production and distribution costs (Greenhut, 1952). Unfortunately, Lösch assumes a homogeneous surface, implying ubiquitous resources with negligible variations in production and procurement costs at alternative locations. Under these conditions, he divides the market so that optimum-sized plants in each industry serve hexagonal sub-areas of equal size and that the plants of different industries serve sub-areas of different size (Lösch, 1954, pp. 124–132). He rotated the system around a common centre, transforming the homogeneous plain into a series of six alternate 'rich' and 'poor' zones. With population and settlement localized in the 'rich' zones, industries became agglomerated in the same zones to enjoy the economies of inter-industry linkage in localized market belts. Lösch shows that the location of plants at the centres of their hexagonal market zones minimizes freight costs on the product and maximizes profit. Isard (1956, p. 272) has redrawn Lösch's original concept to depict more realistic conditions; this modified system comprises increasingly large hexagons (or geometric areas) as distance from an urban agglomeration increases or population density decreases. This emphasizes agglomeration around, rather than in belts leading away from, the node.

The Löschian model is also modified by Greenhut (1952 and 1957) – to embrace cost considerations more thoroughly within an oligopolistic system. Greenhut argues that when firms seek the 'minimax' location (combining

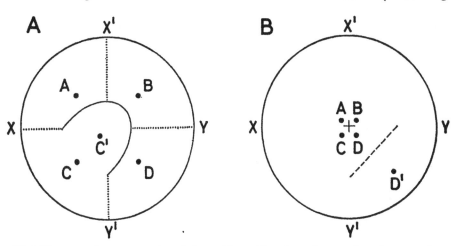

10.4 Diagrammatic representation of the problem of interdependence and maximum profit in an oligopolistic market. Four manufacturers divide a circular market area (A) by locating near the centres of their market segment at A, B, C and D, respectively. Each seeks to control a larger market by making a move to a new location as C does to C¹. Yet because each entrepreneur fears that his market may be reduced in this way, all producers locate near the centre of the total market area (B). This may encourage a new competitor to enter the market and to locate at D¹ to monopolize a sufficiently large market at the expense of D. Manufacturers must weigh up, therefore, the tactics and ability of rivals, as in a game.

minimum cost with maximum profit), the number of variables increases and the doubtful market area expands because 'uncertainty is an inherent part of the capitalist system' (Greenhut, 1957, p. 88). Such a situation arises from the existence of different costs in alternative locations, giving uncertainty about the competitiveness of rivals, and from the profit motive, giving uncertainty about the policies of rivals. Location decisions under these conditions are amenable to the use of game theory. Each competitor adopts one of several possible strategies in production and location choice; chance looms large, therefore, both in the location of individual plants and of industries as aggregates of plants. Further research is required to establish the vagaries of personal and board room decisions in more definite outline and to remove limitations to game theory application (Stevens, 1961) before 'game models' can become valuable geographically. Nevertheless, Greenhut's minimax model demonstrates that uncertainty, rivalry, and oligopolistic competition in undifferentiated products leads to industrial localization where firms think that rivals can achieve optimum sales at lowest cost. Hence man's 'follow-the-crowd' attitude leads to localized specialization. This is especially so in industries that either use fairly 'mobile' materials and manufacture valuable transportable products or serve large markets e.g. motor vehicles, textiles. If the products are differentiated, fragile or bulky, and bear high freight rates the tendency will be towards dispersal to divide the market. The problem of interdependence and maximum profit is shown for an oligopolistic market in Figure 10.4. The difficulty of establishing the minimax location scientifically has been demonstrated by Dunn (1956) who attempted to apply the 'index of location' to find the point in Florida that combines greatest market potential with minimum transport costs. This suggests that most locations are suboptimal or satisficer locations only.

Location and pricing systems

The adoption of different price systems from country to country or from industry to industry can modify resulting location patterns significantly. This section offers, therefore, further embellishment of the basic model. Where delivery prices are fixed equal for a whole market area the entrepreneur will obtain the maximum profit where his costs of selling to the market area are the lowest. This system encourages more firms to agglomerate in the largest market area as compared with other price systems (Greenhut, 1964, p. 387). The single basing-point system tends to encourage and perpetuate industrial concentration at a base point which, as a result, is an artificial least-cost point. The system effectively stifles or reduces the possibility of industrial growth in locations which may have more advantages than the base point (Stocking, 1954). Multiple basing-point systems permit spatial competition from several points so that market areas assume different shapes and sizes; market potentials vary from area to area, therefore, and may induce the selective dispersal of plants. The geographical impact of distance is only allowed full play – within the limits of transport rate policies – with the f.o.b. pricing system. It induces entrepreneurs to divide the market area by dispersing in an attempt to monopolize some segment of it (where demand is elastic); distance is the defence against competition. Again this tendency will be greater the greater the transport costs on the products. The locational relationships are possibly more complex, however, for, as Greenhut (1964, pp. 393–403) demonstrates for the U.S. paint industry, entrepreneurs may obtain their raw materials under one delivery price system but sell their products under another. This seems to be a field in which empirical research is otherwise lacking.

Too little knowledge exists about the geographical effects of business organization in a capitalist society. For instance, the location of wholesale facilities may be under-rated as a factor which 'deviates' the actual market 'centre' from the 'gravitational market centre' and induces the excessive agglomeration of market-oriented industries. This possibility is clearly underlined by Martin (1964, p. 114): 'From the manufacturer's point of view, London's effective demand for a product is disproportionate to its ultimate consumption. The market he supplies directly may consist of wholesalers and intermediaries acting as channels of demand.' Similarly, where demand for a product is fairly scattered, an entrepreneur may map out his potential market area by 'random walk' methods and choose his plant location accordingly.

LOCATION POLICY IN THE SOCIALIST WORLD

One-third of the world's population is living in countries that are managed, as planned economies, by Communist governments. Whether or not these governments apply Marxist-Leninist ideology *per se*, it is well to reflect that they do plan patterns of industrial location which differ significantly from those in 'free enterprise' economies; this is mainly because location is strictly controlled and directed according to specific guiding principles. This does not mean, however, that what is relevant to capitalism is irrelevant to socialism, or vice versa. For although 'western' theorists are described as 'bourgeois apologetic theorists for capitalism' (Balzak et al., 1949, pp. 106–110; Secomski, 1956, pp. 7–20), location planners in the socialist countries do take inspiration from the work of Weber and Lösch (Saushkin, 1959, p. 47; Dziewoński, 1961, p. 7; Mrzygłód, 1962, pp. 42–45). The very facts that Weber tries to prove the raw material-orientation of industry and that Lösch centres his study on market relationships make nonsense of statements that they are trying to construe apologies for capitalist location patterns, especially when Marxist economists and geographers themselves stress (following Marx, 1850) that the distribution of capitalist industry bears little relation either to material sources or to markets. Greenhut (1957, p. 84) even asserts that 'Lösch's rational system could only be brought about by full direction from the state'.

Location under capitalism is criticized by the socialists on the grounds that it leads to disproportions in economic opportunities between regions. The increasing and abnormal agglomeration of industry in 'growth' regions is not counteracted by the disinterest of capitalists and the inadequacy of government planning measures in developing industries in backward, declining, 'stranded' or 'ruined' areas (Hoover, 1947, p. 203) where depleted resources, unemployment, and overspecialization are major problems. Socialist location policy seeks to avoid these 'contradictions' by 'paying most

8

attention, from the very beginning, to the correct location of industry as the decisive element in the development of the whole national economy' (Secomski, 1956, p. 43).

The socialist approach to location differs from the capitalist approach in several important ways. It replaces the individual's profit motive by the State aim at raising production and income to achieve higher, more equal living standards for all everywhere. New industrial development plays the key structural and spatial role in this. The maximum social and economic benefit is required of every project. This involves a 'complex' planning approach to dovetail plant location into the whole system of spatial linkages between a given plant and other industries, settlements, transport facilities, and land uses, with an eye also on local, regional, and national repercussions. Whereas few capitalists compare alternative locations because uncertainties make it difficult for them to do so (Hague and Dunning, 1954, pp. 203–204), long-term plans and projections of future economic and social trends create a higher degree of certainty in the socialist state, so permitting comparative analyses of national, regional, and local needs and potentialities. Such analyses – when subject to defined location criteria – 'create opportunities for a wider choice of location alternatives that permits more flexible distribution (of industry)' (Secomski, 1956, p. 51). Evidence from Yugoslavia suggests that relatively exhaustive comparisons of locations are made (Hamilton, 1962, pp. 146–161; Hamilton, in preparation); an added stimulus in that state is the activation of local interest through polycentric planning (Hamilton, 1964C, p. 181). Decisions made hastily (Hamilton, 1962, pp. 191–201) or made from the viewpoint of individual plants only, as in Poland's Six-Year Plan (Mrzygłód, 1962, p. 25), can cause costly mistakes. In fact, the final location chosen often reflects the outcome of arguments between representatives of the industrial sectors, who stress the most economic location for individual plants in their particular sectors only, and the regional planning representatives whose socio-economic 'complex' approach indicates a different pattern of location as optimal (Lissowski, 1965).

The socialist countries are very varied. The vast space of the USSR is organized in huge economic and administrative regions that are larger in area than the east-central European states. East Germany, Czechoslovakia, and Poland inherited well-developed capitalist industrial regions so that postwar location patterns show considerable *unconformities*. In contrast are the Balkan states where a largely socialist industrial pattern has emerged from the beginning. And China and Yugoslavia are applying diverse planning methods. These differences, however, form the basis for testing, rather than formulating, a model which embodies principles common to all the socialist states.

Four principles of location may be noted. Industries should be located, first, near the sources of the raw materials and fuel they use, and second, near the markets for their products (Livsic, 1947). This is to minimize the transport involved 'from processing the raw materials through all subsequent semi-finishing stages, even to the marketing of the finished product' (Lenin, 1954, p. 33). These principles hint at the clear division between material- and market-oriented industries; yet spatially, the two may be linked, for as Saushkin (1959, p. 49) stresses, 'consumption increases (in mining, fuel, and power areas) as a result of industrialization and the growth of population' while 'the high level of production (in market areas) makes it possible to discover and exploit more efficiently even the second-rate raw materials and fuels'. The bases of industrial 'complexes' or agglomerations emerge, but the social frictions of overcrowding, commuting, and housing needs either encourage the dispersal of interrelated industries within a growing industrial area (Hamilton, 1964D), or prohibit new location in such inherited capitalist agglomerations as Upper Silesia (Pounds, 1959). Moreover, socialist planning removes two powerful forces which encourage agglomeration under capitalism: uncertainty and the need for close contact regarding quick changes in fashion, tastes, and consumer preference.

These considerations operate within the constraints of the third principle: to locate plants to achieve an 'even' distribution of industry (Feigin, 1954, pp. 182–183). An 'even' distribution is not meant to imply some standard ratio of industrial employment to area or to population, although it *could* be understood as that level of industrial development which, combined with other productive activities, ensures equal incomes per head of population over the whole country in the long run. Since incomes are usually unequal, this principle involes the policy 'to develop backward areas' by allocating to them proportionally more new industrial capacity than the national average (Hamilton, 1964B, p. 79). Priority is given to locating those industries in backward areas that can process materials and energy resources locally and employ much labour (Hamilton, 1963, pp. 104–105). Governments undertake geological research in such areas to establish the existence and nature of resources that would strengthen the local bases for processing industry; an example is sulphur in south-eastern Poland (Hamilton, 1964D). The importance of developing backward areas is often overemphasized; in reality the principle is to ensure that 'the development of production in one area would not hinder . . . (that) in other areas' (Saushkin, 1959, p. 49). This rather vague notion is modified and conceptualized by Dziewoński (1958) who writes: 'In view of experience . . . the principle (requires the provision of) equal possibilities . . . and opportunities for social and economic development in all parts of the country and of giving similar living standards'. Naturally a very long time is necessary to achieve this. In fact industrial growth is planned for all regions, including the most developed, to supply capital equipment for factories being constructed in the backward areas in exchange for materials and manufactures lacking elsewhere (Kidrić, 1948, p. 34). Nevertheless, the model involves a progressive dispersal of plants to all regions (defined as 'territorial administrative areas') of the country, and within each region to both the larger and the smaller towns. It also involves balanced industrial employment for both men and women locally, and balanced regional development in stressing variety as well as some degree of specialization often around some heavy industrial core area (Caesar, 1954). The fourth principle is to disperse plants in the interests of security and national defence, as has happened in the location of Yugoslav iron and steel capacities (Hamilton, 1964A, pp. 59–61).

Recently, Mrzygłód developed an industrial location model that is at once amenable to these principles and to those outlined in the 'reformed' Weber model. Socialist planning, asserts Mrzygłód (1962, pp. 35–47), must avoid the two costly extremes of over-concentration and over-dispersion by creating sub-regional industrial complexes. These comprise various plants that use local raw materials, that are 'mobile' and employ local labour and that, as branch or auxiliary plants, are linked with a new or expanded 'parent' industry – or location leader (Estall and Buchanan, 1961, pp. 167–170) – within each area. The number of plants would vary with the economic possibilities (Myrdal, 1957), but they would be assigned to different towns to achieve an even spatial distribution and in such a way as to relate their size to the labour available in each town and town hinterland. A framework of Löschian market areas is replaced by one of labour-supply areas. This would reduce commuting and migration to a minimum. The growth of industry and its effects should ensure a balance between jobs and present and future labour supply *within the region*, except where the exploitation of large and valuable resources requires migration. This implies that priority is given generally to allocating plants to those regions with a labour surplus. Mrzygłód suggests three possible variations on this theme (Fig. 10.5) involving a region A that possesses resources for processing by an industry which is divisible into spatially separated stages and which manufactures a high value product. Transport costs are small. The first case (Fig. 10.5A) assumes that the labour surplus is spread evenly among three adjacent sub-regions. The processes can be divided between these sub-regions, therefore, so that A produces and refines the raw material while sub-regions B and C manufacture semi-finished components and the final product. The second variation (Fig. 10.5B) divides the industry between two sub-regions, A as before, and B which, because of a larger labour surplus, is allocated all the fabricating and finishing stages. The last variation (Fig. 10.5C) assigns the whole industry to sub-region A where the labour surplus is very large; this variation could apply also in the case of an industry using weight-losing materials mined in sub-region A. The character, rate, and spatial arrangement of industrial growth in each sub-region depends upon the needs of the nation as a whole, interacting with the needs and potentialities of the sub-region itself and of the

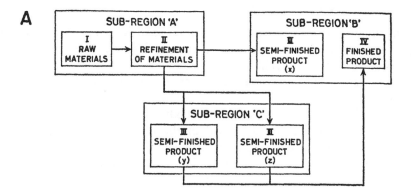

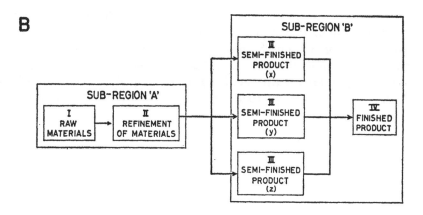

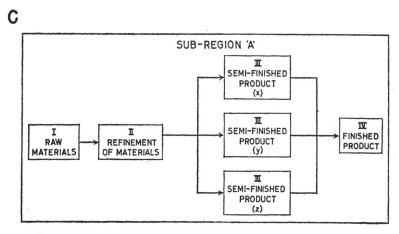

10.5 Alternative plant allocation strategies in a socialist state
(*Source: Mrzygłód, 1962, p. 40*).

adjacent sub-regions. It is obvious that neither spatially nor structurally can 'targets and capacity levels for specific commodities . . . be regarded in isolation' (Ghosh, 1965, p. 91).

The socialist models of general plant location are amenable to the theoretical and – as long as the difficulty of handling a multitude of variables is overcome – also operational, treatment embodied in the assignment or allocation-location models to be discussed in the following section (Silhána, 1964, pp. 321–328). The contributions of the Dutch economists – Tinbergen, Bos and Serck-Hansen – on spatial dispersion are also of value in this respect.

ALLOCATION-LOCATION MODELS

It is necessary to pass on from models of general plant location to consider models that assign particular plants of the same industry or of several industries to a number of locations. Significant here is the use of linear programmes and algorithms to find optimum production locations for a number of plants by allocating optimum capacities either among an equal number of locations or among a wide range of alternative locations. Both conceptual and opera-

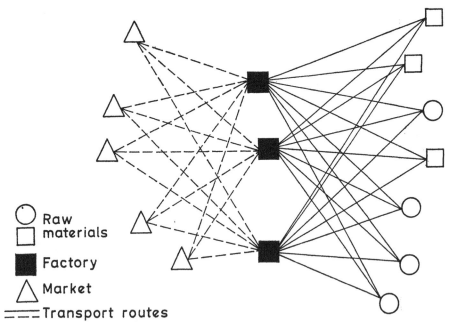

10.6 The Beckmann-Marschak assignment model. This sketches diagrammatically a problem which is to select those material and product flows which will maximize net profit to a corporation with several plants located separately, and to an industry as an aggregate of firms.

tional models are reviewed here. Beckmann and Marschak (1955) assume a diversified environment with many markets, and many potential material suppliers, each with different production costs, and with limited capacity that restricts each plant's size and saleable output. The number of locations being fixed, the task is to allocate optimum *scales* to the plants to minimize the volume and direction of flows and yet to satisfy demand. A macro-economic approach is thus required. The problem is shown diagrammatically in Figure 10.6. The model copes with situations, therefore, where the capacity of plant *A* exceeds the maximum output of its cheapest material's source *x*, so that the balance of supplies must be purchased from a dearer source *y* in so far as sales to the most lucrative market realize maximum profits. Goldman's conceptual model (1958) is complementary and allocates optimum iron and steel capacities among three equidistant islands which each produce one raw material and consume various quantities of steel to minimize both the flows of materials and products and also the backhauls and empty ship movements (social capital). The model also allows variation in the optimum *number* of locations (between 1 and 3), division of producers into separate blast-furnace or steel plants, and the spatial coincidence of the material sources with the markets. A Koopmans transport linear programme proves that optimum locations are a function of total demand (not cost) when transport is provided at marginal-cost rates by ships; plants may be best located, therefore, where transport costs are *above the minimum*. Indeed, the movement of empty transport capacity induces modified location patterns to provide return cargoes – witness the growth of Erie port steel industries and the Urals – Kuznetsk combine (Holzman, 1957).

More complicated are those models which assign *n* manufacturing plants (in one or several industries) to *n* locations to achieve the maximum combined profits (social or planning benefits) from the plants. Koopmans and Beckmann (1957) attack this problem in two ways. First they use the 'linear assignment' approach that treats plants individually; this method allocates plant A to the most profitable location, and then assigns plant B to the most profitable of the remaining locations, and so on. Second, they apply the 'quadratic assignment' approach that counts not only transport between plants but also the effects that the costs and profits of each plant in a given location have upon the locations, costs, and profits of all other plants. The number of variables becomes prohibitive yet the model still 'promises to be an important avenue of approach towards the precise and meaningful analysis of such interdependencies' (Cox, 1965, p. 235). The model could give greater insight into the problems of linkage, agglomeration, and socio-economic costs and benefits. The difficulties which Koopmans and Beckmann found in achieving an optimum solution may be overcome by the choice of a *satisfactory*, rather than an optimal, assignment of plants. Such a solution may be of no less 'optimal' quality (Odhnoff, 1965, p. 39).

Operational linear programme models have been applied in the United States to those industries, notably agricultural processing, in which the number of variables is small. Koch and Snodgrass (1959) construct a model of the optimum geographical distribution of the US tomato-processing industry in relation to its raw materials and markets which assumes the reality of imperfect competition. Transport costs t_{ij}[1] are analysed to include processing and advertising costs, product differentiation, consumer preference for Californian canned tomatoes, and price discrimination in 'secure' markets. This model shows that existing location and production patterns are not optimal. Stollsteimer (1963) evolves a simpler operational model for finding the optimum number, size and location of plants processing one raw material into one product. In practice, entrepreneurs extend the length of the processing season to use capacity more efficiently by processing two or more inputs, as, for example, in the frozen-foods industry (Hiner, 1965). Thus Polopolus (1965) generalizes Stollsteimer's model for a case involving three raw materials and final products (sweet potatoes, tomatoes, and okra), 25 producing origins, and 10 potential processing locations in Louisiana. Costs are adjusted for joint processing of tomatoes and okra; and costs of material supply are estimated for three truck sizes on the basis of present production patterns which are adjusted for the future geographical effects of the continued urban growth of Lafayette, the uneven incidence of pests, and the competition of vegetables with sugar-cane and cotton on farms. One multi-product plant located centrally is the optimum solution for, as plant numbers increase, far higher processing costs result from decreasing scale and offset small savings in assembly costs (Polopolus, 1965, pp. 292–295). Cox (1965), in contrast, approaches the problem chiefly from the marketing angle in a model which uses an algorithm to solve the distribution of motor-vehicle bodies from three factories to four assembly plants. More recently, Efroymson and Ray (1966) have devised a branch-bound algorithm which has been used to solve practical plant allocation problems with upwards of fifty plants. Rapid progress is, therefore, being made.

Nevertheless Cox concludes that linear programme models are defective in that they treat the places of production and consumption as points rather than as areas: whence Polopolus's use of 'origins' of supplies of agricultural produce for manufacture. The danger with linear programme models seems to lie in applying complex methods to oversimplified situations to obtain either obvious or already well-established results. An example is provided by the work of Lefeber (1958, pp. 77–87). This is inevitable, nevertheless, with embryo techniques of analysis. In part, however, difficulties arise in presenting complicated programmes with many stages in published works; much

---

[1] 'T_{ij}' is the algebraic shorthand for the costs of transporting materials or products from the ith sources to the jth destination.

computer programming is a question of laboratory method rather than of conceptualization.

INDUSTRIAL LOCATION AND SETTLEMENT HIERARCHIES

Following Christaller, geographers, until recently, have neglected the relationship of primary and secondary manufacturing to the central place hierarchy. Yet industry was, and in the developing countries still is, the chief vehicle of urbanization. The intensive nature of industrial activity explains the concentration of the labour force and auxiliary services in towns, localizing there consumer and industrial markets which attract other industries and services, and unleashing a process of 'circular and cumulative causation' (Myrdal, 1957, p. 13). It should be possible, therefore, to generalize the relationships between the size, distribution, and types of industry.

The hierarchies of settlements embodied by Christaller and Lösch in their respective models of central places and market areas form a logical framework for the distribution of market-oriented industries. According to Alexandersson's study[1] of American industry only three, out of sixteen, manufacturing industries – construction, printing and publishing, and food-processing – are ubiquitous, i.e. located in all towns with over 10,000 inhabitants and can reflect faithfully the respective regular and irregular urban size-distribution hierarchies of Christaller and Lösch. Employment in these industries is positively correlated with the size of any settlement (Alexander, 1963, pp. 296–299). Individually, the sporadic industries show no such correlation and since collectively they account for most manufacturing, they apparently weaken the hypothesis of complete industrial conformity with Christaller-Lösch systems. Evidence shows, however, that ubiquitous industries stimulate their own growth by generating more income and development locally (Pfouts, 1957; Hirsch, 1959). This can attract sporadic industries to a town when that town attains the threshold (Pred, 1965, p. 168) – the minimum population or the minimum volume of sales required to support those industries. Moreover, different industries require different optimum numbers of plants of different optimum scales to serve the same national or regional market (Bain, 1954); different industries, therefore, supply variously-sized market areas. Ideally, the model proposes that sporadic industrial plants will be located in those settlements that have thresholds that are larger than, or equal to, the plants' desired market areas. This implies that the city rank-size concept is compatible with the idea of an urban-centred hierarchy of market areas (Beckmann,

[1] Alexandersson (1956) found that of 36 broad groups of industry twenty were service activities that were located in every town with over 10,000 inhabitants; only three manufacturing industries were so dispersed, the other 13 being sporadically located.

1957; Pred, 1965). Sporadic activities that are rare in small towns are more frequent in larger towns which tend to support not only larger-scale industries but also a greater range of industries of all sizes as a result of larger population, larger thresholds, and the 'nesting' of a greater number of market areas. A synthesis of the works of Bain (1954), Alexandersson (1956), and Philbrick (1957) – regarding industrial scale, cities, and areal functional organization in the United States – suggests itself here. Accordingly, only the three ubiquitous industries would be located in the smallest towns, while medium-sized centres would contain in addition, food-manufacturing, textiles, leather and shoe, and cement industries. Larger cities would be locations for steel, oil-refining, chemicals, metal containers, and distilling, along with all industries of lower-rank towns. Such products as tyres, rayon, soap, cigarettes and farm machinery appear from factories at the metropolitan level together with all lower-rank industries. The largest-scale activities are localized in the major metropolis – non-ferrous metals, motor vehicles, tractors, and pens – or, with typewriters[1] in the primate city which contains all manufacturing industries.

In practice, some branches of industry are developed in particular locations, not for their accessibility to markets, but for their proximity to bulky or perishable raw materials. Agglomeration at localized sources of raw materials or nodes, especially in a capitalist economy, can provide exceptions to the urban-industrial hierarchy (Haggett, 1965, pp. 135–152). Technical and economic changes have been at work, however, to reduce their importance in the industrial spectrum. Even so, they point the way to two modifications of the model.

The first concerns the general distribution of industry. It involves the principle that manufacturing employment as a proportion of total employment varies directly with the population potential and with local urbanization, but inversely with distance from a metropolitan or major market centre (Duncan, 1959, p. 95). Generally, as urban population densities decline with increasing distance from a central city, there is a more rapid decrease in the intensity of manufacturing and in average plant size (Bogue, 1949). Naturally, the regularity and rates of decline differ with varying regional conditions. Nevertheless, this concept seems to fit reasonably well into Isard's modification of Lösch's market-area system (Isard, 1956, p. 275). However, there tends to be a concentration of large hinterland towns with specialist industries at distances of 30–65 miles from the metropolis. This allows for the 'off-centre' location of large enterprises (e.g. steel, oil-refining, vehicles, typewriters) away from, yet near and obtaining the benefits of the urban centre with a threshold equivalent to their market areas. This tendency is reinforced in a socialist economy by the planned spread of industry to

---

[1] Bain shows that typewriter manufacturing requires the least number of plants to serve the national market (Bain, 1954, p. 36).

achieve more equal regional opportunities. Closely-spaced settlements of relatively similar size and often with related specializations may form a 'dispersed city' (Burton, 1963) which, as a unit, contains industrial plants with market areas which equal the threshold of a single and much larger city. At greater distances from the metropolis (over 250 miles in the U.S.A.), much localized manufacturing can thrive in centres that enjoy monopolistic market conditions under the protection of the prohibitive freight-in costs that arise over long distances.

This modification underlines the probability of a more than proportional localization of sporadic industries in metropolitan centres and also in places where the access to such centres, the frequency and size of other urban centres, and the population potential, are the greatest. The explanation is simple. Ubiquitous industries are small-scale, simple activities that are closely tied to consumer markets and not to other industries (except cans and packaging for food-processing). The thirteen sporadic industries defined by Alexandersson (1956) comprise larger, more complex, and individually more specialized, activities that do demand spatial proximity to each other since 'the bulk of the composite market confronting industry is industry itself' (Kenyon, 1960, p. 168). The point is summarized succinctly by Chinitz and Vernon (1960, p. 130): 'The chain of processes between the raw materials and final products has been growing longer and longer. The tendency for any one plant in the chain to use materials which are already processed has continued to grow. As a result, and in increasing degree, plants hold down their freight-in costs by locating near other plants'. These tendencies in a capitalist system underline the validity of Lösch's 'city-rich' manufacturing belts. For even if his geometric arrangement of areas is unacceptable, the fact remains that sporadic industries (including much mining) grow and perpetuate their growth in the city-rich zones where 'the greatest number of locations coincide, the maximum number of purchases[1] can be made locally, the sum of the minimum distances between industrial locations is least, and in consequence . . . shipments . . . are reduced to a minimum' (Lösch, 1954, p. 24).

The second modification of the basic model concerns the relationship between the different types of industrial activity and the distribution and size of settlements. Rarely do towns of equal size possess equal advantages for the same industries, so that no rules can be laid down for other than ubiquitous activities. Each town specializes according to the natural resources that are available locally, and to the produce that its traders handle. Its economy, however, does not function in isolation; the growth of some kind of manufacturing in one town influences economic trends in neighbouring towns, and vice versa. Curry (1964) shows that the diffusion of industry results in specialization by towns and regions according to the arc-sine law. Industries in

[1] (and sales) – author.

town A determine about one half of the industrial structure of adjacent town B (the remainder being autonomous development), while town C, nearest neighbour of B, has one-third of its structure determined by the industries of A, one-third by those of B, while one-third is autonomous, and so on. The type of industries in towns B, C, . . . n are affected by those in the town of origin A according to the diffusion coefficient $d\text{-}k/t$, in which d and t are units of distance and time respectively. Whence the association and linkage of industries in neighbouring settlements and the evolution of areal specialization. Where the urban network is dense the precise directions of influence of the industry of a given town may reflect the operation of random human processes. In addition, Alexandersson's work shows that where towns are widely spaced there is a tendency throughout the hierarchy towards a wider range of industries; in contrast, where the network of towns is more intense there is a tendency towards a smaller range of industries and greater specialization in each town. This seems to reflect the greater spatial possibilities for inter-urban division of labour, the close spacing of towns offering inter-urban external economies of scale. It may also be explained by the tendency of industries with medium-sized plants (e.g. metal-working, textiles) to 'swarm' (Florence, 1963) in separate, though neighbouring, settlements to gain economies of close linkage. Further support is given, therefore, to the first modification of the model.

Individual branches of industry show varying relationships to settlement size. Mining, quarrying, and forest industries are more typical of smaller, more rural settlements; as the size of settlements increases so the proportion of extractive industries decreases (Winsborough, 1959). Manufacturing has a high index of urbanization, yet medium-sized settlements tend to have a higher proportion of manufacturing employment than larger centres (where services are proportionally more important). Both Winsborough and Duncan distinguish two groups of manufacturing which have divergent locational associations. Processing industries[1] have an urbanization index near to zero, indicating a closer correlation between industrial location and the distribution of the total urban and rural population; their importance shows a negative relationship with increasing settlement size. Fabricating industries[2] show a high index of urbanization and a clearly positive correlation with settlement size. The two extreme distribution patterns in the USA are provided by the scattered textile and furniture industries with low indices, and the highly localized, highly urbanized clothing and motor vehicle industries. Broad support for this model is given by Estall (1966, appendix II) through his index of relative dispersal for industries in New England. While some

---

[1] e.g. Forest products, metallurgy, food-processing, textiles, and other non-durable goods industries.

[2] Especially transport equipment (excluding motor vehicles), printing and publishing, electrical machinery, and metal products.

industries may diverge from this general pattern in other countries, there is no reason to expect any marked deviation.

MODELS OF AN IDEAL SPATIAL DISPERSION OF INDUSTRY

Akin to models of industries framed within the settlement hierarchy are those of spatial dispersion developed by the Dutch economists, Tinbergen, Bos, and Serck-Hansen. These models are treated separately since they are not based upon empiricism but suggest an ideal economic and social cost-benefit distribution of industries, accounting the different optimum sizes of plants and market areas in different industries. Tinbergen (1961) assumes a closed economy in which agricultural production and population are spread evenly. Industries are ranked into a hierarchy according to the number of optimum plants n in each so that $n_1, n_2, n_3 \ldots n_H = 1$. The model does not assign these to locations. By specifying the number of centres and their possible industrial compositions it calculates that combination of plants of various industries in various centres which would minimize production and transport costs (Tinbergen, 1961 and 1964). The hypotheses of the model are that each urban centre with an industry of a given rank h contains also all industries of lower rank, and that only one plant of the highest-ranking

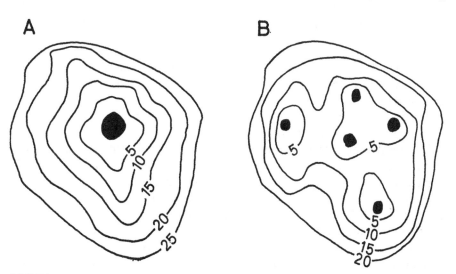

10.7 Dispersion and socio-economic costs. Isodapanes show that the costs of supplying food to one centre where industry is agglomereated may be higher (A) than to several dispersed centres with equivalent demand (B). The concept can be extended to include the social costs, for example, of commuting or of providing extra housing and social facilities in centres of large-scale immigration.

industry is located in any one settlement and that only this plant exports products to other centres and to the agricultural areas.

Bos (1965) modifies this model to include vertically-integrated industries, unevenly distributed population and demand (which are localized in either a large or a small number of settlements), and different settlement types. The aim is to minimize total location costs by dispersing industry to agricultural areas to reduce the costs of supplying agricultural produce to the industrial labour force; its conceptual implications are shown in Figure 10.7. This assumes decentralized marketing of food supplies. In part it suggests Tinbergen's idea (1962) that the full social costs of location (public services, commuting, etc.) should be paid by entrepreneurs to restrict agglomeration to that level at which social benefits equal social costs and to disperse industry so long as the social and economic benefits offset the social and economic costs of doing so. With a large number of centres, industries supply the needs of agricultural areas and they exchange products among themselves. The number of calculations becomes astronomical because plant numbers vary from industry to industry, a very large number of industrial settlement types are possible, and for each type continuous variations in their relative positions are also possible. Bos simplifies by assuming that transport costs to a large number of centres equal those to a circular market area and that plants in each industry have a certain *minimum*, rather than optimum, size. Industries are ranked as in Tinbergen's model but as their number increases the range of possible types of industrial centre increases rapidly. Further simplifications are made, therefore, by omitting the export industry and crosshauls of the same product. Bos concludes (1964, p. 69) that relatively high transport costs per ton-mile for agricultural products encourages decentralization of production over a large number of small centres as the optimum, relatively high transport costs for the products of high-rank industries (those with few plants) concentrate production in a small number of large centres as optimal, and relatively high transport costs for low-rank industry products indicates Tinbergen's system as optimal.

Further refinements are made by developing a simple programming model for several vertically-integrated industries and a small number of settlements. A regularly-shaped area is divided into equal hexagonal market areas (Fig. 10.8). The model indicates that the highest-ranking industries with one plant each will be optimally located in the most central sub-area (1). The regularity of the figure allows lower-rank industries to be located with more freedom since sub-areas 2–7 have equal accessibility to the whole market. The model is adapted here also to show the tendencies of agglomeration at central places (sub-area 1). The unequal number of industries in centres 2–7 represents the impact of unevenly distributed resources and of different optimum locations for different rank industries. It also indicates some social cost disadvantage in sub-areas 2, 4 and 6, and the disadvantage of transport costs in the outer areas (8, 9, 10). Bos stresses that the optimum national location of an industry is *not* only dependent upon transport costs for its own product and inputs, but also upon the costs of transporting the consumer goods required by the working population of the industry. 'For this reason it may *not* be optimal to locate industry 1, delivering its product only to industry 2 ... in the same centre as industry 2' (Bos, 1964, p. 85). Decentralization is thus preferable to agglomeration.

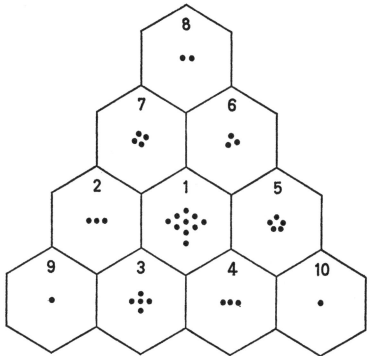

10.8 An idealized model of the spatial dispersion of industry.

Mathematical models of spatial dispersion on similar lines have been developed by Serck-Hansen (1961). This interest in decentralization mirrors the concern of the Dutch for the continuing agglomeration of industry and population in the western Netherlands between Rotterdam and Amsterdam, Tinbergen (1962), for example, asserting that there has been excessive localization of new industry in Rotterdam.

INDUSTRY THROUGH HISTORY

'Manufacture' ante-dates the 'industrial revolution' by many centuries, although most families produced their own clothing, shelter, furniture, and tools. In ancient Rome baking bread, fulling cloth, and making bricks or pottery were large-scale activities (Heaton, 1948, p. 52). A first model might show that, from ancient times until the mid-nineteenth century, the localization of world manufacturing in any era reflected the location of contemporary imperial or dominant power. Areas achieved industrial pre-eminence when they were governed strongly to give internal stability and independence, to exploit local resources and a favourable situation for trade, to expand political power to secure foreign sources of materials, slaves, or skilled labour in order to accumulate capital (Finley, 1965), and to secure adequate food for 'industrial' workers. When these advantages shifted to other areas, the industrial importance of a region waned. The rise and fall of empires and states, explaining the rise and fall of major manufacturing areas, resembles a series of waves rolling slowly across the world through time, from Mesopotamia through Rome and the North Italian city states to England. The process changes and quickens in modern times. A free-for-all is being replaced by civilization bringing the independence and integrity of all states with the knowledge to facilitate and perpetuate the economic and the geographic proliferation of industry. The world is now like a system of differently-sized canal locks or docks, each one a country protected by tariffs and regulations (the weir gates) 'trapping' industrial growth (the water level) to serve home needs (with independent water reserves representing domestic resources), yet inter-connected by channels of trade exchange (inflow and outflow) to balance supply and demand.

A second model might present the intra-regional location pattern before 1700. Capital being limited, processing equipment was usually simple, so that 'industrial' plants were small and *could* be scattered. That they *were* scattered resulted from the interplay of several factors. Energy sources (wood, wind, water, and human or animal muscles) and raw materials in use before 1700 or in the *eotechnic* phase (Mumford, 1934) (wood, agricultural produce, clays, sands, leather, skins, and wool) were ubiquitous. Labour inputs, in volume, time, and skill, were very high, and since the population was also the main market, industries were distributed according to the population and the availability of food. Transport was slow and very costly on land and this limited the area of material or food supply (or of 'putting out') to a maximum radius of a day's journey. Regional differentiation arose chiefly through the growth or lack of growth of industries that processed valuable materials (wool, metals) and produced valuable products (cloth) that were transportable by sea or from a large land area.

The 'industrial revolution' originated in England in the eighteenth century. Adam Smith gives a kaleidoscopic model of its early selective effects. Industrial progress was greater where more advantageous and cheaper transport was available (then water and sea transport). Areas which lacked these facilities, and depended upon land carriage, could develop industry only 'in proportion

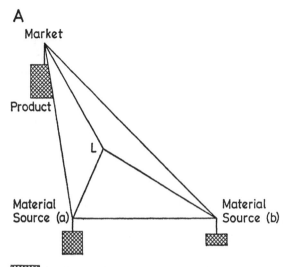

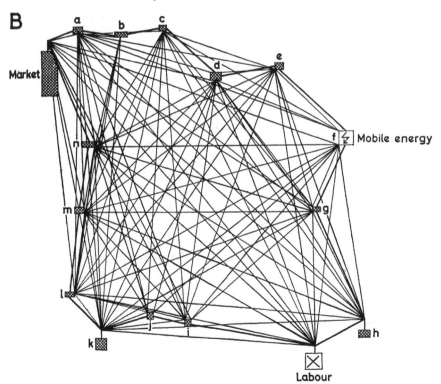

10.9 Location figures: An historical transformation. Weber's location triangle (A) is superseded by a polygon (B) which includes location factors which have become important since 1909.

to (their) riches and populousness' unless they produced goods 'whose price was very considerable in proportion to their weight' (A. Smith, 1776, p. 81). Increasing technical division of labour and improved transport were encouraging areal specialization according 'to the diverse aptitudes, capacities, and resources of different peoples and places' (1808, p. 14). And certain industries were becoming urban activities while others were tending to remain rural activities.

Radical changes occurred in the location of industry as a result of the new technology. Isard (1956) has used a modified Launhardt-Palander model to demonstrate this. The increasing use of coal as a raw material, fuel or power source made possible, and localized, the rapid expansion of existing industries,[1] on or in proximity to the coalfields. Many inventions initiated the growth of new industries[2] largely in the same areas. Weber's model describes the situation well (Fig. 10.9A) since decisions were made by small firms (Goldman, 1958, pp. 91–92) and few industries used more than three raw material inputs. As techniques were wasteful and the major input – coal – was very bulky and partly or wholly 'disappeared' in manufacture (Wrigley, 1961, pp. 3–9) the combined weights of inputs far exceeded the weight of given outputs. The power of the coalfields to localize industry was further strengthened by the occurrence of the initial sources of other raw materials in association with coal, by the expansion of local composite markets as industries and urban populations grew and by the development of infrastructure and skills which encouraged intertia. This model summarizes Mumford's *paleotechnic* phase. Unless Weber's location polygon is treated as an intra-regional figure, therefore, all the weights for many industries pulled on one angle. The location figure was cast into the inter-regional context by 1909, however, as canals and railways 'extended man's mastery over space' (Launhardt, 1885, p. 206) to provide distant locations and nodes with the opportunity to use cheap coal, and as the exhaustion of local resources (other than coal) enforced a dependence upon more distant supply sources. These and other changes were beginning to favour market-orientation (Fig. 10.9A).

Continuous technological progress has shifted the balance more from the materials (especially coal as *the* source of energy) to the markets in the present century; Keir (1921) early recognized this trend in the United States. A modified location figure is presented (Fig. 10.9B) which embodies the key factors of change. These are part of the *neotechnic* phase. The larger size of the figure indicates the ease and economy of linking more distant locations as a result of cheaper and faster transport; the reduction in transport costs (especially for raw materials) has been greater than the reduction in production costs. The smaller symbols for raw materials (k and h) indicate the great reduction in the weight of raw material inputs to product outputs. The

---

[1] These included iron smelting, textiles, pottery, bricks and brewing.
[2] These included metal manufactures, machinery, and transport equipment.

differential costs of moving raw materials are now far less at alternative locations. Other symbols are introduced to denote new location factors. Symbols *a–e, g, i, j, l* and *m* represent the increasing number of intermediate 'pure' manufactured materials (metals, synthetic fibres, plastics), tools, parts, and components, which are important links in the industrial chain. Often of high value or light in weight, individually these are not important location factors, but collectively they may demand the spatial proximity of interlinked plants to each other. The corollary is the growth of diverse finishing and

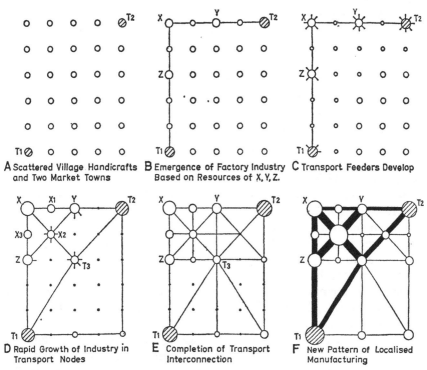

A Scattered Village Handicrafts and Two Market Towns

B Emergence of Factory Industry Based on Resources of X, Y, Z.

C Transport Feeders Develop

D Rapid Growth of Industry in Transport Nodes

E Completion of Transport Interconnection

F New Pattern of Localised Manufacturing

10.10 The emergence of a modern industrial location pattern.

assembly industries which use mainly power and not fuel for heat (Dales, 1953, p. 182) to produce valuable, fragile, bulky, or weight-gaining products, Generally, the freight rates charged for products are high and reflect 'what the traffic will bear'. The increasing complexity of mass-produced articles requires greater contact between producers and consumers and the provision of after-sales service both in the industrial and consumer markets (Chisholm, 1963). For these reasons, the market has become by far the most important single location factor (Fig. 10.9B). A new symbol ('mobile energy') stresses that new energy sources (oil, gas, atomic and ordinary electric power) have made coalfield locations unnecessary, and in consequence, have enhanced the locational importance of market, labour or other locations. The last new symbol (X) represents the growing importance of social factors (personal preference or planning needs), and of labour, particularly the availability of semi-skilled or high skilled personnel; in part, these factors are beginning to underline the importance of locations with pleasant amenities. This trend may be expected to continue as labour becomes more mobile. Automation is both tending to confirm these trends still more (Osborn, 1953) and to increase the *flexibility* of location choices that is indicated by the multitude and range of intersecting lines in the figure.

A model of selective locational industrialization is developed by Pred (1965) to show how progressive agglomeration of manufacturing arose in a few centres in America between 1860 and 1910. Initial advantages emerged at centres where entrepreneurs developed rational or random new industries, because these industries set in train multiplier effects which stimulated invention and attracted immigrants who brought new skills and other industries. Once these centres were thriving, three factors ensured their further growth and discouraged new development in inefficient or non-producing centres: the improvement of transport, the lowering of production costs, and the growth of large combines which could effectively prohibit new entry to an industry by would-be competitors.

The evolution of a modern spatial pattern of industry in a sequence of stages from the handicraft era is suggested in Figure 10.10. Conceptually, the model is an outgrowth of, and an analogy to, the idealized pattern of transport development in an under-developed area that was worked out by Taaffe, Morrill and Gould (1963). A series of evenly-spaced villages (Fig. 10.10A) are locations for small handicrafts in a region which supports two market towns, T_1 and T_2, where more crafts are localized. Fuel and raw material resources are discovered at X, Y. and Z (Fig. 10.10B) which, as a result of new techniques, provide the basis for growth industries. Mines are opened at X, Y and Z to supply materials and fuel to processing industries that develop at T_1 and T_2. Feeder lines develop from all five centres (Fig. 10.10C) and bring competition into the nearest villages, causing handicrafts to contract. The emerging transport network (Fig. 10.10D) intensifies the market area for fuel and minerals causing expansion at X, Y and Z, and for the products of industries located at T_1 and T_2. Manufacturing expands in all these centres as the influx of country people swells their labour forces and consumer markets, and as new industries develop to serve the growing needs of the mines, new factory processes, transport facilities, and urban population. Crafts virtually die (Fig. 10.10D), but some villages become embryo factory locations as transport connections create new nodes (e.g. T_3) and spread the effects of industrial growth around existing nuclei (X_1, X_2, X_3). Existing centres (T_1 and T_2) become more important as the transport network is completed (Fig. 10.10E), but some initial advantages combine with transport advantages to cause greater industrial growth in certain centres. A modified location pattern emerges in Figure 10.10F based on the intensive interlinkage of industries in the resource areas with fine assembly advantages leading to the rapid growth of the most modern industries at X_2, while T_1 maintains its supremacy through inertia and its adaptable labour force. Some theoretically good nodes remain unimportant because of the difficulties of entry for would-be competitors at those locations. A highly differentiated spatial pattern with localized industry therefore replaces the even distribution of activities characteristic of the pre-industrial era.

SOME OBSERVATIONS ON MONTE-CARLO AND MARKOV CHAIN MODELS

The spread of industrial inventions is a promising field of study using Monte-Carlo simulation (Hägerstrand, 1963) and Markov-Chain models (Brown, 1964; Clark, 1965). Certainly the large number of inventions that originated in Britain (and especially in the Midlands and the North of England) after 1700 gives historical confirmation of Hägerstrand's thesis that the amount of additional invention (and thus potential proliferation of industry) is proportional to the amount of existing innovation. Both models stress the prime importance of neighbourhood: an innovation tends to appear near the place where the innovation already exists while the probability of the appearance of the innovation decreases with increasing distance from the existing phenomenon. The models could be applied to trace the diffusion of the stationary steam engine, or any other invention, either within Britain or other countries, or from Britain to regions overseas. This last is the more fascinating problem since it involves certain modifications of Hägerstrand's model. The intensity and the ease of contact and trade (the information field) were far more important than distance in diffusion overseas because resistances from barriers were strong in nearby Europe whereas few barriers were encountered in America. Differences in language, culture, economy, laws, policies, and patents reduced the ability to communicate and to apply ideas and acted as multiple brakes upon invention diffusion in Europe. These barriers clearly delayed the spread of the steam engine from Britain across Europe in both time and distance. Invented by Watt in England in 1768, the efficient steam engine began operating first in Belgium mainly through English enterprise (1780's). It appeared later in France and Germany (1790's), in Austria-Hungary in 1816, and in the Balkans in 1835 where English initiative was again important (Bićanić, 1951, p. 212).

Although they are still in their infancy, Markov-chain models seem to have most potential in tracing trends in industrial location or employment (Williams, 1965) which result, for example, from progressive technological improvement changing scales (and thus distribution patterns) of industrial plants. Markov-chain analysis may also lend itself to problems of industrial migration.

STRUCTURE, PROCESS AND STAGE

Economic growth occurs as agriculture and primary industries decline in importance relative to the growth of secondary and tertiary activities. This implies not only a change in industrial structure through different stages, but also changes in the location pattern. W. M. Davis's cyclical model of landform development suggests itself here as the analogue for a controversial model of industrial structure, process and stage.

Primary activities emerge first in any area. This is the *infancy* stage when industries are pre-eminently extractive and raw material-oriented. The cycle enters *youth* with the growth of manufacturing, especially textiles, to supply the consumer market. *Adolescence* is achieved when basic industries develop to serve producer and consumer markets. Industries tend to become localized near fuel, power and raw material sources and in good nodes as bulk transport becomes important. An area reaches *maturity* when 'it has experienced large-scale development of manufacturing . . . and associated economic development over many decades' and has 'evolved a deep-rooted and highly complex system of industries and services, many of which are inter-related in a variety of important ways' (Estall, 1966, p. 3). Maturity not only suggests a broad and balanced industrial structure, it implies the existence of developed skill and infrastructure embodying the ability to adapt to changing technological and economic conditions by developing regenerative industries – engineering, chemicals, and machine tools or, as in New England, electronics and aircraft (Estall, 1963 and 1966). In Old England no better example can be found than the West Midlands, the cradle of the industrial revolution, where inventiveness has stimulated progressive adaptation to new industries. According to Kenyon (1960, p. 65) the Paterson-Passaic district (New York Metropolitan Area) 'has evolved through several complete cycles since its inception: cotton textiles, locomotive production, silk (and wool) textiles, and lately aircraft manufacture'. An area with a narrow industrial base may find spontaneous adaptation difficult; in this case the area is *immature*. The inception of marked decline indicates *old age* (McCaskill, 1962, pp. 143–169).

Some areas never pass through this cycle. The rise and fall of mining, leaving ghost settlements, is an example of *infant mortality*. Other areas have succeeded in reaching youth with basic industries – for example, the Kielce area of central Poland with metallurgy – when technological or political changes initiated their *premature death*. Nowadays, however, better knowledge and an array of economic and other stimulants (the construction of 'location leaders', communications, or redevelopment) makes industrial *rejuvenation* more likely. Moreover industrializing regions can now progress through the earlier stages of the cycle in several years rather than in several decades. A new stage is now appearing in developed countries where automation in both the secondary and tertiary sectors is initiating a shift again in favour of capital goods.

THE INTERNATIONAL DISTRIBUTION OF INDUSTRY

The world industrial location pattern is uneven and bears no close relation to the distribution of either natural resources (even coal) or population; it can be correlated only with the regional levels of economic development. This suggests a model that is an analogue of Berry's model of developed, developing and underdeveloped countries (in Ginsburg, 1962). Marx (1850) emphasized that localized capital, backed by imperial power, was the most powerful single factor in localizing industry in a small part of the world in the nineteenth century. As inventions revolutionized production and transport, capitalist economy replaced the orientation of industries in Britain to national resources and markets by increasing dependence upon material supplies (except coal) from, and export of products to, the whole world. This suggests a world interaction model, in which capital localization combined with coal in the 'mother' country to localize there industry which depended for resources or markets upon areas (colonies) which were denied manufacturing. The rise of new independent states began to duplicate the process, but industry remains confined largely to former imperial and to capital- and resource-rich countries. As a result, industry is very highly localized in the northern hemisphere, the world manufacturing belt representing a diamond that is oriented from east to west and that is split well to the west of centre by the North Atlantic Ocean.

In detail the world industrial pattern today bears the marked imprint of the nation protecting its own industries and markets by tariffs. This is promoting the international deglomeration of industry, while attempts towards greater international integration are tending to work in the opposite direction. Large states offer large markets for large-scale industries and in the absence of particular planning controls, except where huge distances make regional autonomy an economic proposition as in the U.S.S.R., industries tend to localize. In contrast, they are dispersed in a similar geographical area that is politically fragmented. Figure 10.11 demonstrates this contrast with reference to the textile industry in the U.S.A. (with bordering areas of Canada) and in a comparable area of Europe. Greater economic development (arising from enlarged markets and exchange among integrated territories) according to

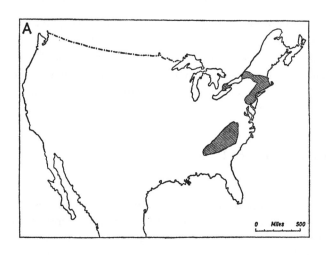

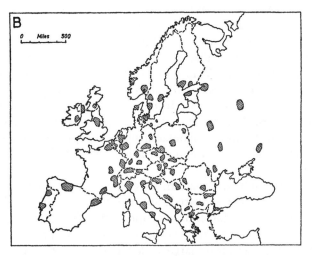

10.11 The comparative effects of political unity and of fragmentation upon industrial distributions: textile industrial areas in the United States (A) and a comparable area of Europe (B).

Myrdal's 'spread effects' means more industries and more plants that *can* be located more evenly. In practice, greater integration calls forth large-scale activities which tend, if uncontrolled, to cause greater regional differentiation than before by localizing in those places which are the most accessible to the enlarged market and which provide most linkages for modern industries. Integration in the European Economic Community, therefore, is consolidating industrial employment (Chisholm, 1962, p. 10) in a 'T-shaped complex of industrial areas that is likely to reap the greatest benefits from Common market conditions' (Wise, 1963, pp. 135–136). This is confirmed by Leviquin (1962) who shows that most of the 800 new enterprises that were developed in the European Economic Community between 1959 and 1961 were located in existing industrial agglomerations. Where economic development is planned, as in COMECON, integration can accelerate the economic 'take-off' of backward areas. The development of new industrial zones along the Vistula waterway in Poland provides an example.

INDUSTRY IN THE NATIONAL SETTING

Theorists from von Thünen to Tinbergen have framed location models in the abstract context of the isolated state. The assumption here, however, is an independent nation which has limited resources and which engages in trade to secure imports of deficient foods, materials, and manufactures. Energy and raw material resources are unevenly distributed within the country, being more contiguous than coincident with the better agricultural areas. In response, the distributions of population, settlement and transport facilities are uneven, tending to be concentrated where mineralized and good agricultural zones overlap or adjoin. Land use becomes less intensive with increasing distance from such concentrations. Given these assumptions, a number of general industrial patterns emerge in all nations.

High transport costs make perishable, bulky, or low value materials and products 'immobile' (Tinbergen, 1962), so that food processing, timber, building materials, paper, and packaging industries are located in most nations and in some regions within each nation. Sporadic industries develop to process localized domestic resources provided that suitable types and quantities of energy (which also condition industrial structure) are available (Dales, 1953). Otherwise, distributions accord with the principles established earlier and with the regional 'climaxes' to be discussed. Every country contains areas with little industry which contrast with those where sporadic industries localize to share common links or facilities, to use associated resources, and to serve associated composite markets. The nation's level of economic development will condition the size, extent, and intensity of such localizations through the degree of inter-industry integration of vertical, diagonal, convergent and indirect kinds (Florence, 1948); the transport pattern conditions their shape and alignment. Models of the shapes and sizes of sample 'manufacturing belts' are given in Figure 10.12. Two special types of industrial centre – the metropolis and the ports – may lie within, on the periphery of, or near this belt and extend or intensify its economic attraction, or they may be located well away from it and act as 'counter-magnetic poles of growth'. Whereas eccentrically-situated import/export-oriented industries

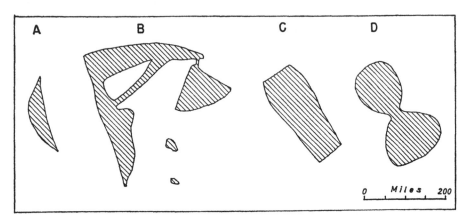

10.12 Sample manufacturing belts.
A: The 'crescent moon' in the western Netherlands, comprising an area between Eindhoven, Rotterdam and Amsterdam.
B: The German 'gallows', including the north–south Rhine axis, the west–east Ruhr–Magdeburg axis and the triangular area of industrial concentration in Saxony.
C and D: The manufacturing belt of England represented as a 'coffin' and an 'hour-glass'.

do develop at the ports where there is a large break of bulk, they do not develop at the land frontiers (of either seaboard or land-locked states) where the same transport facilities continue with no break of bulk. Strategic considerations may also discourage such frontier development. Yet border locations will attract industry if resources are localized there (e.g. coal in Western Europe) or if differential tariffs operating along the frontier offer opportunities for serving two or more national markets instead of one (e.g. the Maggi and Knorr food-processing combines along the German-Swiss frontier).

REGIONAL INDUSTRIAL 'CLIMAXES'

The reality of intricate spatial interdependencies lies at the root of an increasing emphasis in geography upon 'complex' economic regions, central place studies, and city regions. Presented here are some complementary stochastic models of regions with different spatial structures of industry. Hitherto this problem has received most attention in studies of the 'territorial-production complex' in the Soviet Union, chiefly by Kolossovsky, Saushkin, Alampiev, and Pokshishevsky. The original idea occurs in Weber's work (Friedrich, 1929, p. 196) and is developed broadly by Chardonnet (1953). Recently Isard has applied the notion to an oil-refinery, petrochemicals, and synthetic fibres complex in Puerto Rico (Isard *et al.*, 1959). The models that follow are constructed on the premiss that, given the capital and the market demand, differences in regional industrial patterns are a function of the character, size, and variety of available natural resources and of human resourcefulness. They resemble Kolossovsky's eight 'energy-material-manufacturing' cycles (Kolossovsky, 1958). These are modified, however, to encompass variations in population distribution, labour supply, and composite market demand. None should be interpreted rigidly, either in structure or in spatial distinctiveness (Lonsdale, 1965). Rather they suggest regional 'climaxes' of sporadic industries that can be modified by external factors in the same way that vegetational 'climaxes' can be.

The simplest model depicts the scattered or localized extraction of valuable minerals in sparsely-populated regions which suffer from isolation and harsh physical environment. Only localized concentrating, refining, and power plants are important. The second model comprises temperate forest areas where dispersed saw mills and more localized pulp-paper, chemical distillation, and veneer factories are typical; or tropical areas with pharmaceutical, chemical, and rubber industries. In either variant, plants tend to localize in forest margins between the material sources and the markets. These regional models may overlap with each other and be coincident also with sources of cheap hydro-electric power and an electro-metallurgical/electro-chemical complex. If regional demand is large enough, specialist machinery industries may develop within any of these 'climaxes' to meet their demands.

The next two models relate to predominantly farming areas, although 'islands', 'inliers', or 'outliers' of other mining and manufacturing complexes may occur within them. The fourth model, for a commercial livestock region, comprises: widely scattered dairies, more localized leather tanning, meat-packing, and woollen textiles industries; and labour-intensive textile, engineering, or specialist manufactures where labour surpluses exist. The spatial structure of industry in the fifth model of a commercial mixed-farming area is more intricate, as an 'organism' of inter-related plants. Scattered factories process food crops, cotton, tobacco, and oil-seeds. Other industries develop alongside processing plants to use by-products (e.g. for livestock-feed), to supply the processing industries with cans, boxes, machinery from more localized plants, to serve farming needs in fertilizers, wire, tools, machines, and tractors (from plants that are centrally located to one or more farm regions), to serve the farm community in household goods, and to

employ female labour in textile, footwear, printing, and other light industries.

The most dominantly industrial landscapes are associated with the bitu-
minous coalfields and their related metal and non-metallic mineral resources.
The sixth model, then, comprises a complex of coal-mining, coke-chemicals,
gas-chemicals, electric power, heavy metallurgical, glass, and cement industries; the bulkiness of materials and of the products also ties heavy engineering industries closely to the coalfields. Less localized are industries using
lighter by-products, or power from the coalfield, and employing female
labour – plastics, artificial fibres, and textiles. Large fields give rise to major
urban concentrations and transport networks, so attracting consumer-
oriented and transport equipment (e.g. locomotives and ships) to nodes around
the coalfields. Somewhat similar complexes can develop in regions with large
ore-fields. Small or brown coalfields attract few industries other than electric
power, gas and chemicals. The seventh model is typified by two variable
structures: first, iron-ore mining, iron and steel production and manufacture,
with coke-chemicals and second, non-ferrous metal-mining and processing
with sulphuric acid and fertilizer chemicals integrated with smelters to use
by-product gases. Oil-drilling and refining is the chief component of the
eighth regional model which comprises highly localized petro-chemicals and
synthetic fibres plants, associated with decentralized oil and chemical processing equipment, textile-processing, and power industries. Gas may be an
important energy source for other manufacturing industries (Kortus, 1963).

The last two models concern areas where human resourcefulness is most
important: the ports and the metropolis. Large ports have a triple spatial
structure. Industries that process bulky, divisible, imported materials or
weight-gaining products[1] are located along the waterfront. Industries that
either serve, or use the products of, the first group are located near the port.[2]
Light industries that are labour-oriented are located in the suburbs. The
metropolis has an industrial structure which responds to its administrative,
educational, and commercial functions, large labour supply and large and
varied composite market: clothing, luxury goods, office and business
machinery, scientific instruments, vehicles, electrical equipment, light
chemicals, furniture, pharmaceuticals, and foods. If the metropolis is large,
then substantial cost differentials exist between alternative locations within
the city, especially with regard to land, labour, and transport. The model of
the metropolitan spatial structure of industry (Fig. 10.13), therefore, comprises differing localizations of associated industries in different optimum
locations (Haig, 1927). Central locations are occupied by industries (Group A

[1] For example, grain-milling, sugar-refining, public utilities, oil-refining, metallurgy,
cables, vehicles, shipbuilding, and heavy engineering and machinery.
[2] These include food manufacture, light chemicals, furniture, newspapers, engineering, cans,
packaging, and machinery.

in Fig. 10.13) in which the need for the best access to skilled labour from the whole area (e.g. instruments, tools, printing), to the central business district (e.g. clothing, office machinery), and to the whole urban market for distribution (e.g. services, newspapers), offsets high land costs (Stefaniak, 1963; Lowenstein, 1963). The 'swarming' of closely associated activities in small

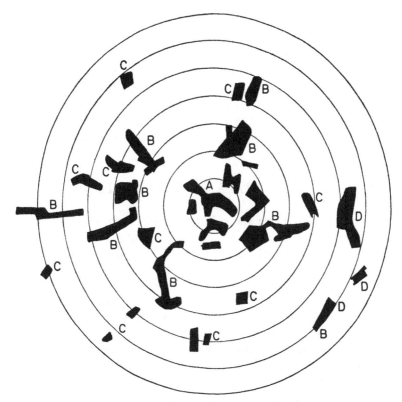

10.13 A model of the spatial industrial structure of a metropolis. This model is based on a map of the industrial areas of London in Martin (*1964, p. 122*). The radii are at intervals of 2½ miles. This should be compared with Lowenstein's (*1963*) synthesis of the intra-urban industrial patterns of sample American cities. Note the 'wedging' effect of transport lines on the distribution pattern, offering greater possibilities in the inner or middle ring for external economies and linkage.

enterprises gaining external scale economies in the central district may give rise to sharply-defined industrial quarters, e.g. the clustering of metal plating, printing machinery and parts, printing works, typesetting, lithographic plate-making and ancillary trades in the Clerkenwell–Fleet Street area of London (Martin and Hall, in Clayton, 1964, p. 34). Larger enterprises seek cheaper land, a good location for material assembly and product distribution, and access to unskilled or semi-skilled male and female suburban labour. They locate along radial or 'ring' transport arteries and include port industries (Group B in Fig. 10.13), food manufactures, electrical, engineering and light industries (Group C). On the outskirts are industries which either require large land areas for assembly-line production, for stores and for waste, or which are dangerous and obnoxious; this group (D) comprises vehicles, heavy engineering, oil-refining and heavy chemicals, metallurgical and paper industries.

Chardonnet suggests the existence of three other regional complexes: autarkic, colonial and strategic (Chardonnet, 1953, pp. 168–182).

Model urban centres possess all, or some combination of, industries appropriate to their hierarchical importance in the region in which they are situated. 'Exceptional' centres occur where random industries develop, or where two or more regional types coincide or adjoin. Set theory and Venn diagrams may be used to clarify the typological 'position' of any city.

INDUSTRIAL INERTIA AND MIGRATION

Industrial patterns are never static, but most areas provide examples of both industrial *inertia* and *migration* (Costa Santos, 1961, pp. 7–9). A model of *inertia* demonstrates that the advantages of present location far outweigh the advantages of relocation through high fixed costs of plant, the existence of a large skilled labour nucleus which facilitates high quality output and the application of new techniques *in situ*, the local development of good infrastructural facilities, and the existence of complex interlinkages with other local industries that makes the movement even of the more 'footloose' industries a major operational cost risk. 'The power of a locality to hold an industry . . . greatly exceeds its original power to attract. The new locality must not only excel the old, but it must excel it by margin enough to more than offset the resisting power of the matrix' (Ross, 1896, p. 265).

While several models of population migration have been developed (Haggett, 1965, pp. 35–40), industrial counterparts are virtually lacking. The main reason lies in the different 'migration' processes involved. Rarely does industrial migration, unlike population migration, involve a physical movement from one area to another. Shifts in plant location were partly of this character in the U.S.S.R. during the Second World War, and in Yugoslavia after the war (Hamilton, 1963, p. 103). Usually 'migration' results from differential rates of industrial growth which is accentuated if stagnant or declining and expanding industries are localized in separate areas. The exhaustion of some resource may initiate the contraction of an industry in one area, while the existence of that resource elsewhere may stimulate the growth of the industry in another area. Some traditional coalfield industries, particularly iron and steel, migrated to other areas as long-distance import of low-grade materials (which were formerly available in the coalfield area) became necessary. Technological change is a major cause of migration. Once barren of industry, many coalfield areas became major industrial regions, while manufactures in forest and upland areas declined. General technological progress – at once encouraging concentration and dispersion (Gravier, 1954) – has progressively freed industry from the tight locational bonds of the water-power site or coal-mine. Canals and railways permitted industrial growth at greater distances from mines, yet they still tied factories to the canal bank, waterfront, or railway siding. In the twentieth century motor transport and electricity transmission have further unfrozen industrial concentrations. A 'centrifugal-dispersion' model is implicated here; it works against the 'centripetal-concentration' model that is induced by the economies of scale and integration.

The development of branch plants plays an important role in migration and reflects the economies that accrue in marketing expanded production from, applying new techniques in, or obtaining a stable labour force at, a

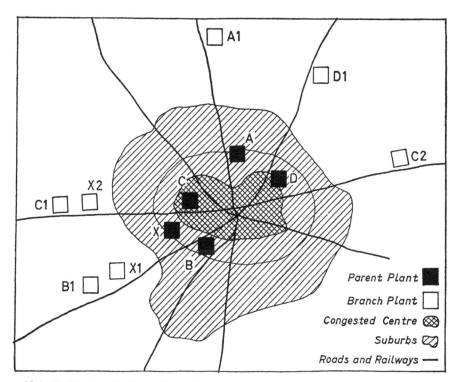

10.14 Radial migration from a large city. Branch plants are established in clearly radial fashion outside the city by firms with parent plants within the city. The foundation of a branch C2 by C is given as an example of what is *unlikely* to happen as a result of inconvenience in maintaining contact.

separate location (McLaughlin and Robock, 1949; Greenhut and Colberg, 1962). Analyses of industrial change in and around Greater London (Martin, 1964 and 1966; Keeble, 1965) show that the establishment of branch plants outside, by firms with parent plants inside the conurbation contributes significantly to centrifugal movement. This suggests a 'radial migration' model (Fig. 10.14) in which inner or outer suburban firms construct branches outside the conurbation which are accessible by the line of 'least transport effort'. This involves the shortest and most direct exit through the suburbs; branches are rarely located where links with parent plant require transport across the congested central areas. Such 'migration' may also be part of the 'suburbanization' of industry (Scott, 1963). In detail the process is more complex than these models imply. Luttrell (1962) shows how variations in transport, administrative and overhead costs between parent and branch plants condition the type of branch plant that is developed at a given distance from the main factory. A model can be constructed to distinguish three types of 'industrial distance-migration'. Branches that manufacture components for parent plants in 'footloose' industries (which serve a national market) are located, for easy control, within a radius (in Britain) of 60 miles of the main factory usually to tap available labour supplies. Intermediate plants, with some separate output, are less dependent upon the main plant and are located within a radius of 150 miles of it. Completely self-contained branch plants are established at greater distances, especially to supply growing regional markets more efficiently.

Much more research is needed before satisfactory models of inter-regional industrial migration can be presented. It seems that such models must embrace: the spatial effects of technological change; the 'regional elasticity' of industrial growth, which depends upon the importance of first, the expansion of existing plants as compared with the construction of new factories and second, the large-scale integrated plants and 'immobile' industries as compared with smaller scale or divisible plants and 'footloose' industries; comparative regional trends in population, labour and market conditions and in resource availability; and the inter-regional effects of these changes. Fuchs (1959) considers that the chief component should be the differences between regional industrial structures. Vanhove (1961), however, attaches importance to regional labour reserves, wage levels and industrial diversification, while Garrison (1960) suggests a three-region industrial migration model using Lövgren's input-output technique (1957). Markov-chain models, possibly when combined with input-output analysis, may offer a solution of the problem.

INTRA-REGIONAL CHANGE AND THE MULTIPLIER MODEL

Every factory movement and expansion sets in motion a chain reaction of corresponding scale that can be embodied in a 'multiplier' model. A useful conceptual and operational tool, the multiplier model can demonstrate the direct and indirect regional economic effects of the construction of a new, or closing of an old and important industry and the overall and selective effects on industry of general economic and social changes. The model provides an interesting example of usually positive direct and indirect relationships with positive feedback, although there may be negative side-effects where substitution is involved. Barfød (1938) showed that, by closing an oil factory employing 1,300 workers in Aarhus, some 6–10,000 workers in industries and services would become unemployed in and around the city. The total and selective regional growth effects of the location of a steel combine have been calculated by Isard and Kuenne (1953) for the New York-Philadelphia area. Pred's multiplier model (1965, p. 165) is adapted to describe graphically the long-term local impact of this project (Fig. 10.15). The same total or same

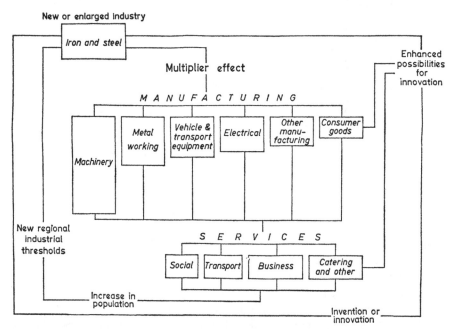

10.15 The multiplier effects of the location of an integrated steel plant in the New York-Philadelphia area. No time period is laid down for the achievement of these changes. The increases in employment (shown proportionately by the size of each box) in each activity were calculated for six rounds of chain expansions.

particular impact should not be expected in another environment, especially where capital is more scarce and less mobile and where industry and transport are less developed. The model emphasizes the importance of location leaders in shaping regional industrial structure giving greater substance, therefore, to the models of regional industrial complexes suggested earlier.

Broad socio-economic changes also induce differential changes in various industries. Hirsch (1959) asserts that a $1 million increase in final demand in the St. Louis Metropolitan area would induce four times more growth in

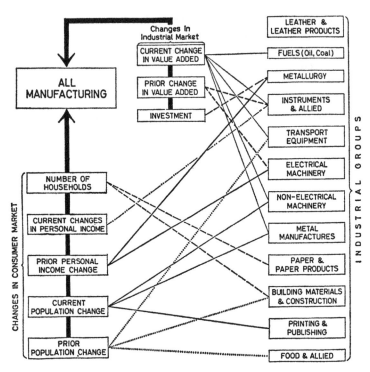

10.16 The more important relationships between eight types of change in intra-regional demand and thirteen manufacturing industries.

employment and sale turnover in printing as in petroleum industries. By using input-output techniques he demonstrates that activities with high direct employment changes per unit investment (e.g. textiles, transport equipment, printing) have also high indirect effects, while growth in timber, furniture, paper, machinery and miscellaneous manufacturing has relatively small effects on employment. Like Hirsch, Mattila and Thompson (1960) consider that intra-regional trends are more important vectors of regional change than are extra-regional factors. They analyse the effects of eight prior and present changes in State consumer and industrial markets (Fig. 10.16) upon 20 American manufacturing industries. Thirteen of these industries were strongly influenced by these changes. Figure 10.16 summarizes schematically the type of industry upon which each kind of market trend had the greatest influence. Clearly current changes in value added, for example, cause the greatest changes in the development of fuel, instruments, transport equipment, non-electrical machinery and metal manufacturing industries while changes in the number of households influence the paper and building materials industries most. Each industry is affected positively by the eight market factors with the exception of leather which seeks locations where labour supplies are readily available in existing or declining market areas. The diagram does not attempt to show more than the major effects of change upon industry, nor does it signify inter-industry relationships.

AN INDUSTRIAL EXAMPLE: IRON AND STEEL

Throughout this essay frequent references have been made to the model location affinities of various branches of processing and manufacturing industry.[1] It seems appropriate at this stage to refer to one industry in greater detail. The iron and steel industry has been chosen partly because it has become almost a model for industrial location in general, chiefly on account of the importance of changing production techniques and its apparent suitability to Weberian analysis. However, the industry is typical for other reasons. Its

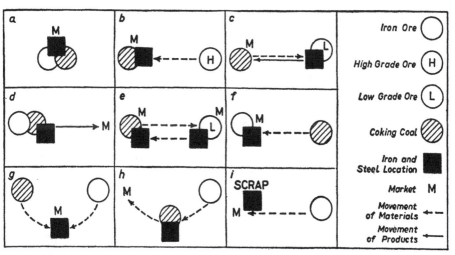

10.17 Model locations of iron and steel industries.

[1] Scale models are not in question here, for these, at best, are suitable for presenting site relationships and plant layout.

necessity to each and every national economy has involved the establishment, in a variety of locations, of various types of integrated iron and steelworks and non-integrated steel plants using a variety of processing techniques to cope with the variable inputs that are available in grade and mix. With the wide range in qualities of iron ores and coals that can be used to produce iron and, with scrap-steel and with the choice of Bessemer, Siemens-Martin, Kaldo, Tysland-Hole, electric arc and L.D. processes of steel-making, the model for the industry clearly stresses that there is no single optimum location where transport and other costs combined are at a minimum. As Moses states, 'if inputs are substitutible . . . three problems are . . . inseparable: optimum output; the optimum combination of inputs; and optimum location' (Moses, 1958, p. 272). The range of 'model' locations that can be optima is summarized diagrammatically in Figure 10.17. To these may be added the less frequent but no less viable location of non-integrated steel plants where abundant hydro-electric power or labour supplies exist, or, as location leaders, where regions are in need of development; some plants, too, may be strategically located in isolated areas (Hamilton, 1964A, pp. 62–63). These 'social' decisions may sometimes involve integrated plants, their development having been encouraged by national pride or by the activity of strongly organized regional interests. As in reality, then, the model must recognize the operation and viability of industry in sub-optimal or in satisficer locations. In this respect the iron and steel industry is more typical of the bulk of manufacturing industry than it is of industries which must be located in close proximity to their raw material supply sources.

CONCLUSION

Certain threads in model-building emerge from the complex weave of the industrial location pattern. The variety of approaches to the subject is abundantly clear. Scale is important in conditioning the degree of factual abstraction and the selection of location factors to be incorporated in the models. The variables to be handled differ as between larger and smaller scales of generalizing the organization, development and classification of industry and between different scales of spatial analysis and activity interdependence. Through time there has been a shift of emphasis which is reflected in the contrast between the models of Launhardt and Weber and those of plant assignment and spatial industrial dispersion. Such a shift is expressed in the increasing importance of macro-economic, at the expense of micro-economic, location decisions. It underlines the fact that the location decision is no longer the last act in the decision-taking process. The need to develop a given industry in a particular region generates feedback which influences the choice of the scale of plant and the nature of its output. The iron and steel industry provides a good example. The change of emphasis has been facilitated by the evolution of more refined mathematical techniques. Equally significant is the increasing importance of descriptive and stochastic models at the expense of deterministic models, especially in analysing the spontaneous evolution of the modern industrial pattern. Models of random processes and diffusion are appropriate so long as industrial location is not strictly and scientifically planned. Game theory, despite its limitation (Stevens, 1961), is applicable to the realm of oligopolistic capitalism. As industrial location becomes a matter of national and social, as well as economic interest, problems of plant assignment and dispersion take on an enhanced value. To be comprehensive, such models must embrace factors which, for lack of data, cannot be reduced at present to a common money denominator. Nevertheless, the complexity of plant allocation models – even when they use only cost and profit data – is such that 'the model-builder is torn between the ends of reality and manageability' (Koch and Snodgrass, 1959, p. 155). The latter problem is being overcome by the use of high-speed computers; but the former will remain a problem of correct programming. Many of the models that have been discussed in this essay have sought optimum location solutions – for individual plants, for an industry and for plants and industries in the framework of total spatial, social and economic interdependencies. In the light of the imperfections of reality they should seek satisficer locations. Yet the quest for optimum locations at all levels is valuable as a means of discovering alternative satisfactory solutions. As far as mortals are concerned, therefore, models that satisfy may also optimize.

REFERENCES

ALEXANDER, J. W., [1963], *Economic Geography*; (Englewood Cliffs), 661 pp.
ALEXANDERSSON, G., [1956], *The Industrial Structure of American Cities*, (Lincoln, Nebraska), 134 pp.
ALKJAER, E., [1953], *Erhvervslivets Beliggenhedsproblemer*, (Copenhagen), 90 pp.
BAIN, J. S., [1954], Economies of Scale, Concentration and the Condition of Entry in Twenty Manufacturing Industries; *American Economic Review*, 44, 15–39.
BALZAK, S. S., VASYUTIN, V. F. and FEIGIN, YA. G., [1949], *Economic Geography of the U.S.S.R.*; ed. by Harris, C. D., (New York), 620 pp.
BARFØD, B., [1938], *Local Economic Effects of a Large Scale Industrial Undertaking*, (Copenhagen), 74 pp.
BECKMANN, M., [1957–58], City Hierarchies and the Distribution of City Size; *Economic Development and Cultural Change*, 6, 243–248.
BECKMANN, M. J. and MARSCHAK, T., [1955], An Activity Analysis Approach to Location Theory; *Proceedings Second Symposium on Linear Programming*, (Washington), 331–379.

BIĆANIĆ, R., [1951], *Doba Manufakture u Hrvatskoj i Slavoniji 1750–1860*, (Zagreb), 459 p.

BOGUE, D. J., [1949], *The Structure of the Metropolitan Community: A Study of Dominance and Subdominance*; (Ann Arbor), 210 p.

BOS, H. C., [1965], *Spatial Dispersion of Economic Activity*, (Rotterdam), 106 pp.

BROWN, L., [1964], *The Diffusion of Innovation: A Markov Chain-Type Approach*; Northwestern University: Department of Geography, Discussion Paper No. 3.

CAESAR, A. A. L., [1955], On the Economic Organisation of Eastern Europe; *Geographical Journal*, 121, 451–469.

CHARDONNET, J., [1953], *Les Grands Types de Complexes Industriels*, (Paris), 196 pp.

CHINITZ, B. and VERNON, R., [1960], Changing Forces in Industrial Location; *Harvard Business Review*, 38 (1), 126–136.

CHISHOLM, M., [1962], The Common Market and British Manufacturing and Transport; *Journal of Town Planning Institute*, 1–12.

CHISHOLM, M., [1963], Tendencies in Agricultural Specialisation and Regional Concentration of Industry; *Papers and Proceedings, Regional Science Association*, 10, 157–162.

CLARK, W. A. V., [1965], Markov-Chain Analysis in Geography: An Application to the Movement of Rental Housing Areas; *Annals of the Association American Geographers*, 55, 351–359.

CLAYTON, K. M., (Ed.), [1964], *Guide to the London Excursions*, (London), 162 pp.

COSTA SANTOS, M., [1961], *Descentralizaçao Industriãl no Estado de São Paulo*, (São Paulo), 74 pp.

COTTERILL, C. H., [1950], *Industrial Plant Location: Its Application to Zinc Smelting*, (St. Louis), 155 pp.

COX, K. R., [1965], The Application of Linear Programming to Geographic Problems; *Tijdschrift voor Economische en Sociale Geographie*, 56 (6), 228–236.

CURRY, L., [1964], The Random Spatial Economy: An Exploration in Settlement Theory; *Annals of the Association of American Geographers*, 54, 138–146.

DALES, J. H., [1953], Fuel, Power and Industrial Development in Central Canada; *American Economic Review*, 43, 181–198.

DEAN, W. H., [1938], *The Theory of the Geographic Location of Economic Activities*, (Ann Arbor).

DEVLETOGLOU, N. E., [1965], A Dissenting View of Duopoly and Spatial Competition; *Economica*, 32 (126), 146–160.

DUNCAN, B., [1959], Population Distribution and Manufacturing Activity; The Non-Metropolitan United States in 1950, *Papers and Proceedings, Regional Science Association*, 5, 95–103.

DUNN, E. S., [1956], The Market Potential Concept and the Analysis of Location; *Papers and Proceedings, Regional Science Association*, 2, 183–194.

DZIEWOŃSKI, K., Translator and Editor, [1961], *Gospardarka Przestrzenna*; (Economics of Location, by A. Lösch), Warsaw, 407 pp.

EFROYMSON, M. A. and RAY, T. L., [1966], A branch-bound algorithm for plant location; *Operations Research*, Baltimore, 3, 361–368.

ESTALL, R. C., [1963], The Electronic Products Industry of New England; *Economic Geography*, 39, 189–216.

ESTALL, R. C., [1966], *New England: A Study in Industrial Adjustment*, (London), 296 pp.

ESTALL, R. C. and BUCHANAN, R. O., [1961], *Industrial Activity and Economic Geography*, (London), 232 p.

FEIGIN, YA. G., [1954], Razmeshchenie Proizvoditel'stva pri Kapitalizmie i Socializmie; *Izvestia Ekonomiki Akademii Nauk SSSR*, (Moscow), 552 pp.

FINLEY, M. I., [1965], Technical Innovation and Economic Progress in the Ancient World; *Economic History Review*, 18 (1), 29–45.

FLORENCE, P. S., [1948], *Investment, Location, and Size of Plant*, (London), 211 pp.

FLORENCE, P. S., [1953], *The Logic of British and American Industry*, (London), 368 pp.

FOLEY, D. L., [1956], Factors in the Location of Administrative Offices; *Papers and Proceedings, Regional Science Association*, 2, 318–326.

FRIEDRICH, C., [1929], *Alfred Weber's Theory of the Location of Industries*, (Chicago), 256 pp.

FUCHS, V. R., [1959], Changes in the Location of U.S. Manufacturing since 1929; *Journal of Regional Science*, 1 (2), 1–18.

FULTON, M. and HOCH, L. C., [1959], Transportation Factors Affecting Location Decisions; *Economic Geography*, 35, 51–59.

GARRISON, W., [1959–60], Spatial Structure of the Economy; *Annals of the Association of American Geographers*, 49, 471–482, and 50, 357–373.

GHOSH, A., [1965], *Efficiency in Location and Interregional Flows*, (Amsterdam), 95 pp.

GINSBURG, N., [1960], *Essays on Geography and Economic Development*; University of Chicago, Department of Geography, Research Paper, No. 62

GLEAVE, M. B., [1965], Some Contrasts in the English Brick-making industry; *Tijdschrift voor Economische en Sociale Geographie*, 56 (2), 54–62.

GOLDMAN, T. A., [1958], Efficient Transportation and Industrial Location; *Papers and Proceedings, Regional Science Assocation*, 91–106.

GORUPIĆ, D., [1954], Ekonømski Obračun Investicionog Programa; *Ekonomski Pregled*, (Zagreb) 2, 421–446.

GOSS, A., [1962], *British Industry and Town Planning*, (London), 190 pp.

GRAVIER, J. F., [1954], *Décentralisation et Progrès Technique*, (Paris), 387 pp.

GREENHUT, M. L., [1952], Integrating Leading Theories of Plant Location; *Southern Economic Journal*, 18, 526–538.

GREENHUT, M. L., [1956], *Plant Location in Theory and Practice*, New (York), 338 pp.

GREENHUT, M. L., [1957], Games, Capitalism and General Economic Theory; *The Manchester School of Economic and Social Studies*, 25, 61–88.

GREENHUT, M. L., [1960], Size of Markets versus Transport Costs in Industrial Location Surveys and Theory; *Juornal of Industrial Economica*, 8, 172–184.

GREENHUT, M. L. and COLBERG, M. R., [1962], *Factors in the Location of Florida Industry*, (Tallahalasee), 108 pp.

GREENHUT, M. L. and WHITMAN, W. T., (Eds.), [1964], *Essays in Southern Economic Development*, (North Carolina), 385–403.

HÄGERSTRAND, T., [1963], *On the Monte-Carlo Simulation of Diffusion*, (Mimeographed).

HAGGETT, P., [1965], *Locational Analysis in Human Geography*, (London), 339 pp.

HAGUE, D. C. and DUNNING, J. H., [1954], Costs in Alternative Locations – The Radio Industry; *Review of Economic Studies*, 22 (3), 203–213.

HAIG, R. M., [1927], Major Economic Factors in Metropolitan Growth and Arrangement; *Regional Plan of New York and its Environs*, (New York).

HAMILTON, F. E. I., [1962], Regional Changes in the Location of Industry in Yugoslavia; Unpublished Ph.D. Thesis, University of London, 338 pp.

HAMILTON, F. E. I., [1963], The Changing Pattern of Yugoslavia's Manufacturing Industry; *Tijdschrift voor Economische en Sociale Geographie*, 54 (4), 96–106.

HAMILTON, F. E. I., [1964A], Location Factors in the Yugoslav Iron and Steel Industry; *Economic Geography*, 40, 46–65.

HAMILTON, F. E. I., [1964B], The Skopje Disaster; *Tijdschrift voor Economische en Sociale Geographie*, 55 (3), 78–80.

HAMILTON, F. E. I., (Ed.), [1964C], *Abstracts of Papers*; 20th International Geographical Congress (London), 361 pp.

HAMILTON, F. E. I., [1964D], Geological Research, Planning and Economic Development in Poland; *Tijdschrift voor Economische en Sociale Geograpie*, 55, (12), 251–253.

HAMILTON, F. E. I., [in preparation], *Yugoslavia: Patterns of Economic Activity*; (London).

HARRIS, C. D., [1954], The Market as a Factor in the Localisation of Industry in the U.S.; *Annals of Association of American Geographers*, 44, 315–348.

HEATON, H., [1948], *Economic History of Europe*, (New York), 792 pp.

HINER, O. S., [1965], Economic Developments in the Grimsby-Immingham Area; *Tijdschrift voor Economische en Sociale Geographie*, 56 (1), 21–32.

HIRSCH, W. Z., [1959], Inter-industry Relations of a Metropolitan Area; *Review of Economics and Statistics*, 41, 360–369.

HITCHCOCK, F. L., [1941], The Distribution of a Product from Several Sources to Numerous Localities; *Journal of Mathematics and Physics*, 20, 224–230.

HOLZMAN, F. D., [1957], The Soviet Urals-Kuznetsk Combine: A Study in Investment Criteria and Industrialisation Policies; *Quarterly Journal of Economics* 71, 368–405.

HOOVER, E. M., [1937], *Location Theory and the Shoe and Leather Industries*, (Cambridge, Mass.), 323 pp.

HOOVER, E. M., [1948], *The Location of Economic Activity*, (New York), 310 pp.

HOTELLING, H., [1929], Stability in Competition; *Economic Journal*, 39, 41–57.

ISARD, W., [1956], *Location and Space Economy*, (New York), 350 pp.

ISARD, W. and KUENNE, R. E., [1953], The Impact of Steel upon the Greater New York-Philadelphia Industrial Region: A Study in Agglommeration Projection; *Review of Economics and Statistics*, 35, 289–301.

ISARD, W. and others, [1960], *Methods of Regional Analysis*, (Cambridge, Mass.), 784 pp.

ISARD, W., SCHOOLER, E. W. and VIETORISZ, T., [1959], *Industrial Complex Analysis and Regional Development*, (Cambridge, Mass.), 294 pp.

KATONA, G. and MORGAN, J. N., [1952], The Quantitative Study of the Factors Determining Business Decisions; *Quarterly Journal of Economics*, 46, 67–90.

KEEBLE, D. E., [1965], Industrial migration from north-west London, 1940–64; *Urban Studies*, 2, 15–32.

KEIR, M., [1921], Economic Factors in the Location of Manufacturing Industry; *Annals American Academy of Political and Social Science.*

KENYON, J. B., [1960], *Industrial Localization and Metropolitan Growth: The Paterson-Passaic District*; University of Chicago, Department of Geography Research Paper No. 67, 224 pp.

KIDRIĆ, B., [1948], *Odnosi Izmedju Narodne i Privredne Politike*, (Belgrade), 146 pp.

KLEMME, R. T., [1959], Regional Analysis as a Business Tool; *Papers and Proceedings, Regional Science Association*, 5, 71–77.

KOCH, A. R. and SNODGRASS, M. M., [1959], Linear Programming Applied to Location and Product-Flow Determination in the U.S. Tomato-Processing Industry; *Papers and Proceedings, Regional Science Association*, 5, 151–166.

KOLOSSOVSKY, N. N., [1958], Proizvodstvenno-Territorial'nie Sochetanie (Kompleks) v Sovetskoi Ekonomicheskoi Geografii; *Osnovi Ekonomicheskogo Raionarovania*, (Moscow), 200 p.

KOOPMANS, T. C. and BECKMANN, M., [1957], Assignment Problems and the Location of Economic Activities; *Econometrica*, 25, 53–76.

KORTUS, B., [1963], Kompleks Przemysłowy Apszeron; *Przegląd Geograficzny*, 36 (4), 569–590.

KRYŻANOWSKI, W., [1927], Review of Literature on the Location of Industries; *Journal of Political Economy*, 35, 278–291.

KUHN, H. W. and KUENNE, R. E., [1962], An Efficient Algorithm for the Numerical Solution of the Generalized Weber Problem in Spatial Economics; *Papers and Proceedings, Regional Science Association*, 8, 21–33.

LAUNHARDT, W., [1882], Die Bestimmung des zweckmässigsten Standorts einer gewerblichen Anlage; *Zeitschrift des Vereins Deutscher Ingenieure*, 106–115.

LAUNHARDT, W., [1885], *Mathematische Begründung der Volkwirtschaftslehre*, (Leipzig), 218 pp.

LEFEBER, L., [1958], *Allocation in Space*, (Amsterdam), 151 pp.

LENIN, V. I., [1954], *Dzieła*, (Warsaw), 3 vols.

LEVIQUIN, M., [1962], *Marché commum et localisations*, (Paris-Louvain).

LINDBERGH, O., [1953], Economic-Geographical Study of the Localisation of the Swedish Paper Industry; *Geografiska Annaler*, 35, 29–49.

LISSOWSKI, W., [1965], Wpływ Układ Działowo-Gałęziowego na Układ Regionalny Planu Perspektywicznego, *Biuletyn Komitetu Przestrzennego Zagospodarowania Kraju* (Warsaw), No. 36.

LIVSIC, R. S., [1947], Nekotorie Teoriticheskie Voprosy Razmeshchenia Promishl'ennosti; *Izvestia Akademii Nauk SSSR*, Otdiel Ekonomiki i Prava, 4.

LONSDALE, R. E., [1965], The Soviet Concept of the Territorial-Production Complex; *American Slavic Review*, 24 (2), 466–478.

LORIA, A., [1888], Intorno della influenza della rendita fondaria sulla distribuzione delle industrie, *Accademia dei Lincei, Rediconti*, 4, 114–126.

LORIA., [1898], Ricerche Ulteriori della influenza della rendita fondaria sulla distribuzione delle industrie; *Rediconti*, 14, 235–243.

LÖSCH, A., [1954], *The Economics of Location*, (New Haven), 520 pp.

LÖVGREN, E., [1957], Mutual Relations Between Migration Fields: A Circulation Analysis; *Proceedings, Symposium on Migration in Sweden, Lund Studies in Geography, Series B, Human Geography*, 13, 159–169.

LOWENSTEIN, L. K., [1963], The Location of Urban Land Uses; *Land Economics*, 39, 407–420.

LUTTRELL, W. F., [1962], *Factory Location and Industrial Movement*, (London), 2 vols.

MANNERS, G., [1962], Some Location Principles of Thermal Electricity Generation, *Journal of Industrial Economics*, 10 (3), 218–230.

MANNERS, G., [1965], Areas of Economic Stress – The British Case; in Wood, W. D. and Thomas, R. S., (Eds.), *Areas of Economic Stress in Canada*, (Kingston, Ontario), 221 pp.

MARCH, J. G. and SIMON, H. A., [1958], *Organisations*, (New York).

MARTIN, J. E., [1964], The Industrial Geography of Greater London; in Clayton, R., (Ed.), *The Geography of Greater London*, (London), 111–142.

MARTIN, J. E., [1966], *Greater London: An Industrial Geography*, (London), 292 pp.

MARX, K., [1850], *Manifesto of the Communist Party*, (London), 48 pp.

MARX, K., [1909], *Capital: A Critique of Political Economy*, (Chicago), 3 vols.

MATTILA, J. M., and THOMPSON, W. R., [1960], The Role of the Product Market in State Industrial Development; *Papers and Proceedings, Regional Science Association*, 6, 87–96.

MCCASKILL, M., (Ed.), [1962], *Land Livelihood: Geographical Essays in Honour of George Jobberns*, (Christchurch), 280 pp.

MCLAUGHLIN, G. E., and ROBOCK, S., [1949], *Why Industry Moves South*; (National Planning Association, Committee of the South), 148 pp.

MOSES, L. N., [1958], Location and the Theory of Production; *Quarterly Journal of Economics*, 72, 259–272.

MRZYGŁÓD, T., [1962], *Polityka Rozmieszczenia Przemysłu w Polsce, 1946–1980*, (Warsaw), 279 pp.

MUMFORD, L., [1934], *Technics and Civlization*, (London), 495 pp.

MYRDAL, G., [1957], *Economic Theory and Underdeveloped Countries*, (London), 168 pp.

ODELL, P. R., [1963], *An Economic Geography of Oil*, (London), 219 pp.

ODHNOFF, J., [1965], On the Techniques of Optimizing and Satisficing; *Swedish Journal of Economics*, 67 (1), 24–39.

OSBORN, D. G., [1953], *Geographical Features of the Automation of Industry*; University of Chicago, Department of Geography, Research Paper No. 30, 106 pp.

PALANDER, T., [1935], *Beiträge zur Standortstheorie*, (Uppsala), 419 pp.

PFOUTS, R. W., [1957], An Empirical Testing of the Economic Base Theory; *Journal American Institute Planners*, 23, 64–69.

PHILBRICK, A. K., [1957], Principles of Areal Functional Organisation in Regional Human Geography; *Economic Geography*, 33, 299–336.

POLOPOLUS, L., [1965], Optimum Plant Numbers and Locations for Multiple Product Processing; *Journal of Farm Economics*, 47, (2), 287–295.

POUNDS, N. J. G., [1959], Planning in the Upper Silesian Industrial District; *Journal of Central European Affairs*, 18, 409–422.

PRED, A., [1965], Industrialisation, Initial Advantage, and American Metropolitan Growth; *Geographical Review*, 40 (2), 158–185.

QUINN, J. A., [1943], The Hypothesis of Median Location; *American Sociological Review*, 8, 148–156.

ROSCHER, W., [1899], *Nationalökonomik des Handels und Gewerefleisses*, (Berlin), 730 pp.

ROSS, E. A., [1896], The Location of Industries, *Quarterly Journal of Economics*, 10, 247–268.

SAUSHKIN, YU. G., [1959], *Economic Geography of the U.S.S.R.*, (Oslo), 148 pp.

SCHÄFFLE, G. F., [1878], *Bau und Leben des sozialen Korpers*, 3, (Tübingen).

SCOTT, D. R., [1963], The Location of Metropolitan Industries; *Western Australia*, 6 (4), 31–41.

SECOMSKI, K., [1956], *Wstęp do Teorii Rozmieszczenia Sil Wytwórczych*, (Warsaw), 135 pp.

SERCK-HANSEN, J., [1961], *Some Mathematical Models on the Spatial Distribution of Industry*, (Rotterdam).

SILHÁNA, V., (Ed.), [1964], *Ekonomika Průmyslu Č.S.S.R.*; (Prague), 657 pp.

SLETMO, G., [1963], *Geographical Distribution of Economic Activity in Norway*, (Bergen), 15 pp.

SMITH, A., [1776], *The Wealth of Nations*, (London), 3 vols.

SMITH, A., [1808], *The Economists Refuted*, (London).

SMITH, D. M., [1966], A Theoretical Framework for Geographical Studies of Industrial Location; *Economic Geography*, 42 (2), 95–113.

SMITH, W., [1952], *Geography and the Location of Industry*; (Liverpool), 20 pp.

SMITH, W., [1955], The Location of Industry; *Transactions, Institute of British Geographers*, 21, 1–18.

SMITHIES, A., [1941], Optimum Location in Spatial Competition; *Journal of Political Economy*, 49, 423–439.

SOMBART, W., [1919], *Der Moderne Kapitalismus*, 2 vols.

STEFANIAK, N. J., [1963], A Refinement of Haig's Theory; *Land Economics*, 4, 428–433.

STEVENS, B. H., [1961], An Application of Game Theory to a Problem in Location Strategy; *Papers and Proceedings, Regional Science Association*, 7, 143–158.

STOCKING, G. W., [1954], *Basing Point Pricing and Regional Development*, (North Carolina), 274 pp.

STOLLSTEIMER, J. F., [1963], A Working Model for Plant Numbers and Locations; *Journal of Farm Economics*, 45, 631–645.

TAAFFE, E. J., MORRILL, R. L. and GOULD, P. R., [1963], Transport Expansion in Underdeveloped countries: A Comparative Analysis; *Geographical Review*, 53, 503–529.

TINBERGEN, J., [1961], The Spatial Dispersion of Production: A Hypothesis; *Schweizerische Zeitschrift*, 97 (4), 1–15.

TINBERGEN, J., [1962], Research on the Geographical Decentralisation of Industry in the Netherlands; in *Guest Lectures in Economics*, Henderson, E. and Spaventa, L., (Eds.), (Milan), 230–242.

TINBERGEN, J., [1964], Sur une modele de la dispersion géographique de l'activité économique; *Revue d'économie politique*, 74 (1), 30–44.

VANHOVE, N. D., [1961], *De Doelmatigheid van het Regionaal-Economisch Beleid in Nederland*, (Eeklo, Belgium), 157 pp.

WALLACE, L. T. and RUTTAN, V. W., [1960], The Role of the Community as a Factor in Industrial Location; *Papers and Proceedings, Regional Science Association*, 6, 133–142.

WEBER, A., [1909], *Über den Standort der Industrien, I: Reine Theorie des Standorts*, (Tübingen), 246 pp.

WEBER, A., [1914], Industrielle Standortslehre; *Grundiss der Sozialökonomik*, 4.

WILLIAMS, S. W., [1965], The Changing Character of the Fluid Milk Processing Industry in Illinois; *Illinois Agricultural Economics*, 32–39.

WINSBOROUGH, H. H., [1959], Variations in Industrial Composition with City Size; *Papers and Proceedings, Regional Science Association*, 5, 121–131.

WISE, M. J., [1963], The Common Market and the Changing Geography of Europe; *Geography*, 48 (2), 129–138.

WRIGLEY, E. A., [1961], *Industrial Growth and Population Change*, (Cambridge), 193 pp.

CHAPTER ELEVEN

Models of Agricultural Activity

JANET D. HENSHALL

A GENERAL MODEL

The study of agriculture has interested some of the most able members of our profession. Pioneer works by such men as O. E. Baker, Olaf Jonasson, Clarence F. Jones, Samuel van Valkenburg and Griffith Taylor published in *Economic Geography* in the inter-war period have been seen as the greatest contribution of that journal to our subject (Buchanan, 1959, p. 6). These studies of the agricultural regions of the world established the broad pattern, and they were followed by numerous empirical papers on agricultural land use analysing the unique causes of patterns within a specific area. According to the most recent overviews of agricultural geography (McCarty, 1954; Buchanan, 1959; Reeds, 1964) we have advanced little in our methods during the forty years since Baker's paper of 1926. The urgent need for more research is recognized by our famine frightened world, yet in 1964 it could still be said, 'Agricultural geography has not yet advanced beyond a primitive stage of development simply because many studies have been superficial investigations of extensive areas' (Reeds, 1964, p. 52). If we are to make the leap forward into maturity we must take a fresh look at our data, concepts and methods.

In the past work was often limited by a shortage of published data. Today there is a vast amount of detailed information available for most parts of the world but Coppock's (1964B) *Agricultural Atlas of England and Wales* is one of the very few examples of a geographical study utilizing this data bank. Recent work in theoretical geography has influenced our conceptual approach to agricultural problems. These developments have three major aspects: firstly there is a more theoretical approach and less concern with the 'uniqueness' of geographical distributions (Bunge, 1962); secondly there is a retreat from the deterministic interpretation of phenomena to a probabilistic and behaviourist one; thirdly the micro-geographic study is assuming a greater importance (Blaut, 1959; Brookfield, 1964). Finally our methods have benefited from the 'quantitative revolution' of the last decade and in particular the availability of computers able to handle large amounts of data has opened up new research possibilities. The wind of change in this branch of our discipline has brought with it a new awareness of the ideas other disciplines have to offer and communication with the related subjects of pedology, agroclimatology, plant ecology, rural sociology and agricultural economics in particular has increased. This chapter aims to analyse the present trends in agricultural geography with particular reference to the theoretical models developed.

The study of agriculture is concerned with individual farms having certain characteristics of area, soil, crops, livestock, etc. and complicated functional relationships based on the natural environment, the agricultural economy and the rural society. In general systems terms (von Bertalanffy, 1951) we are dealing with a set of objects (farms) with attributes (characteristics) functionally related through circulating movements (of money, labour, etc.) with energy inputs in response to the social and biological needs of the system. This system contains sub-systems, for example the plantation system, which form our agricultural regions or farming-type areas.

Models represent simplified parts of our system. As our information field becomes more complicated we turn to models for aid in understanding reality. Models of agricultural activity may be divided primarily according to whether they are based on a part of the system, that is on a farm, or on the total system or sub-system. Our experimental and analogue models are usually of the first type whilst models of agricultural land use are of the second type. Some models are *normative* in their approach (describing what ought to be under certain assumptions) whilst others are *descriptive* (describing what is actually existing). We may further divide our models on the basis of their treatment of the problem, that is whether the emphasis is on the economic (way of earning a living) or behavioural (way of life) aspects of agriculture. In practice the economic models so far developed tend to be normative, whereas the behavioural models tend to be descriptive. But the framework of the model does not define its content and some models now straddle the class boundaries.

EXPERIMENTAL MODELS

Many geographers have seen the individual farm as the focal point of our studies and the basis of our understanding of an agricultural region (Platt, 1930, 1942; Birch, 1954; Blaut, 1959; Keuning, 1964). The farm is the 'black box' of agricultural activity and the 'field' of our inquiry.

Model farms

The toy model farm with its moveable elements is probably the first model of geographic activity met by a child and as such can play an important role in his education. In the Junior School the model becomes dynamic and the toy is replaced by visits to a real farm (Farm Adoption Scheme). In this way the young child learns about the basic processes of agricultural activity. In the Secondary School childish things are put away and such models disappear from the curriculum although the Ministry of Education (1960, p. 23) does suggest the study of a single farm as an exercise in field work.

There are many ways in which model farms can be used by geographers for as 'a fundamental unit of resource utilization and organization' (Blaut, 1959, p. 81) it is a valuable tool in the task of describing and analysing the patterns of agricultural activity. Heller (1964) suggests that it may be used to describe regions, to study seasonal variations in agricultural activity and to examine the relationship between the environment and human institutions. Blaut (1953) in his study of a one-acre farm in Singapore used this actual farm as a model of resource use. The agronomist has long used experimental farms as aids to research but in the 1950's when the Imperial College of Tropical Agriculture in Trinidad set up a series of experimental 'peasant' farms they were used to illustrate a wide range of economic and social problems as well as the agronomic problems.

Developed from the 'case-study' approach was the 'representative farm' used by economists in the 1920's for type-of-farming studies. As more statistical data has become available this early method has become less popular, although it can still be useful. Stamp (1948, p. 348) suggests that in order to understand the relationship between the land use pattern and the farmer's resources, decisions, abilities, and adaptation to changes in technology and price levels, the land use map should be accompanied by a parallel study of a specimen farm. In the tropics the pattern of land use is complicated by intricate mosaics of plants and multiple crop seasons. In order to understand this pattern it has been found helpful to borrow techniques from the botanist and look not only at the individual farm but at the arrangement of plants within an individual field. On small farms in the tropics the land use pattern is made up of characteristic crop combinations often with a common spatial arrangement amongst the elements of the association, as for example, the ubiquitous hedges of pigeon pea around the fields of sugar cane in Barbados. In these cases the study of a specimen farm is essential to an understanding of the basic pattern of land use (Innis, 1961; Henshall, 1964). Occasionally such micro-study may be used to illustrate changes in land use. Beeley (1965) by mapping individual fruit trees on holdings in southern Turkey is able to show the trend away from subsistence crop mixtures towards the orderly pattern of commercial citrus orchards.

Theoretical model farms

The agricultural economists in their search for the representative farm have largely abandoned the specific farm unit. The modern tendency is to define a composite or hypothetical farm which is in some sense representative of the population being studied. The type of hypothetical farm constructed depends to some extent on the statistical information available. Often the modal farm in the frequency distribution of farms from the same universe is chosen. Butler (1960) uses this method in his study of farms and smallholdings in the North Riding of Yorkshire. The area was selected because of its apparent homogeneity but in fact revealed a variety of farming systems when a modal type was defined and the deviations from the mode examined. The modal farm was based on crop/livestock combinations as shown by the Ministry of

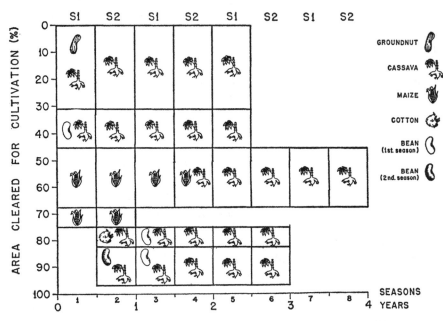

11.1 Model of the agricultural system in the *savane sableuse* of Kasai, Congo (*Source: Beguin, 1964*).

Agriculture returns and supplemented by visits to farms in different size groups. Using this approach Butler showed that a basic differentiating factor in this supposedly uniform agricultural region was the farm operator's attitude to technological innovations. In this case the use of a modal farm was very useful in illuminating the research problem.

Representative farms are a useful method of bringing micro-data to bear on macro-problems but they do involve an aggregation problem (Day, 1963). Carter (1963) criticizes modal farm studies because they are static in nature whilst the farm firm is operating in a dynamic framework, and he feels the model is most useful in studies of low income farms and highly specialized types of farming. Heller (1964) constructs model farms based on averages and includes both farm site and economic features. He suggests that these models may be used for testing the logical bases of agricultural location and for analysing shifts in agricultural patterns over longer time periods than can usually be dealt with on the basis of actual farms. In this way the overall pattern is seen in terms of its basic unit and thus some of the underlying relationships may be perceived more clearly.

Where the average farm is defined on the basis of detailed field work the background knowledge of the research worker may make the model even more useful. Johnson (1964) used his field data to define an 'average' farm and then he used this model as a basis for policy making. Beguin (1964) uses his knowledge of the Congo to develop a model illustrating tropical rotation systems (Fig. 11.1).

In an area where shifting agriculture is practised the farm is no longer the basic unit but is replaced by the area cultivated in the cycle of agriculture, as in Beguin's model (Fig. 11.1). Only rotations occupying more than 5 per cent of the area are shown. Reading horizontally the figure shows the succession of plants in the rotation whilst the abscissa shows the importance of the rotations in the total area cultivated. This type of model is an important element in the understanding of any land use pattern in the tropics.

CONCEPTUAL MODELS

Type-of-farming models

There are two main farming types for which theoretical models have been developed and used extensively: plantation and peasant. Generally these models have been derived by anthropologists or sociologists and economists and they reflect the approaches of these disciplines. The geographer has tended to accept these models as given and there has been little effort to reduce the noise level for geographical problems. But recently we have become aware that the traditional models are no longer applicable to contemporary agriculture and have begun to define new models (Gregor, 1965; Franklin, 1962 and 1965).

The characteristics considered typical of the plantation are crop and areal specialization, highly rationalized cultivation and harvesting techniques, large operating units, management centralization, labour specialization, massive production, and heavy capital investment. Perhaps the oldest explanation of the plantation sees it as a solution to the inability of the white man to do manual labour in the tropics. Research has shown since the last war that Europeans can work as well in the tropics as elsewhere but the plantation is still associated with tropical crops and monoculture. Industrialization (Waibel, 1941) and a large permanent labour force with little mechanization (Buchanan, 1958) were held to be qualities of the model limiting it to the tropics.

Gregor (1965, p. 221) in a very comprehensive study of the modern plantation has said that 'the usual classification of plantation farming as a tropical institution can no longer hold in the face of continued agricultural rationalization'. He notes that many of the characteristics formerly assumed to be the prerogative of tropical plantations are becoming associated with farms in extratropical areas. In those areas traditionally thought of as dominated by the plantation, monoculture has declined and cooler zone crops have gained in importance. The economic basis has changed and markets are no longer always in the metropolitan country or even foreign, but may be domestic as in the case of the Brazilian sugar plantations. Entrepreneurs are ethnically more diverse but may belong to the same group as the workers, and the ownership of a plantation may now be in the hands of an individual, a corporation or the state. Hutchinson (1959, p. 38) describes the modern Brazilian plantation system as 'a constellation of the corporate relations of the planter-industrialist and the private planter-supplier'. This corporate plantation he calls the 'new' plantation. Labour on the plantation is now less plentiful but has more in common with the industrial worker in its adherence to union rules than to the agricultural worker. The demographic aspects of the model have been developed by T. Lynn Smith (1959) who suggests that the population of plantation areas has certain characteristics more closely related to those of urban populations than rural populations although this has been challenged by Pico (1959) with reference to Puerto Rico and Henshall (1966) in the case of Barbados.

As the plantation has spread and changed many new types and regional variations have been recognized. New typologies have been set up such as that of Gerling (1954) based on processing complexity, or Steward (1960) based on historical stages of development and cultural variation, or Wolf (1959) identifying differential cultural adaptation of labour to the modern plantation society. The old plantation model has been replaced by a dynamic model which recognizes a continuum of change and development.

Models of peasant agriculture fall into two groups; those which use the 'way of life' approach, usually set up by anthropologists, and those which use the 'way of earning a living' approach, usually preferred by economists. As an example of the first approach Firth (1951) uses the word 'peasant' to describe any society of small producers for their own consumption, and Robert Redfield (1956, p. 18) describes peasants as people whose 'agriculture is a livelihood and a way of life, not a business for profit'. Wolf (1954) stresses that a peasant lives on land he controls and that an agriculturist who carries on agriculture for business and reinvestment, looking on the land as capital and commodity is a farmer not a peasant. Elena Padilla (1960) defines a peasant as 'an organizational type characterized by individual ownership of the land or undivided rights over the productive unit, family and kin labour, and the use of a simple technology to raise cash crops in addition to subsistence crops' (Padilla, 1960, p. 25).

The economists approach the problem in a different way. Edwards and Rees (1964, p. 73) give us a useful working description of our model listing peasant characteristics as follows: 'small scale of operation; heavy reliance on human labour provided mainly by the peasant and members of the family, and assisted in some systems by animal and mechanical power; use of traditional ("backward") techniques and a strongly conservative attitude towards innovation; individual rather than co-operative or collective cultivation of land; and a significant concentration on production for home consumption.' Anne Martin (1958, p. 88) defines the peasant as a farmer with little or no education and therefore resistant to change and technologica innovation.

Our basic model seems to describe a small-scale agricultural producer dependent on family labour using simple methods but Franklin (1962) points out that this commonly accepted view of peasant life is full of paradoxes. We have for example the idea that the peasant is a good and careful farmer yet in reality his yields are often very low; peasant family life is extolled for its virtues yet many peasants will sacrifice their own lives to educate their children to leave the system; village solidarity with its appeal to the romantic is realistically compared to class or caste differences that often present strong barriers to community development. Franklin (1962, p. 3) shows that 'a fundamental cause of division within the village lies in the inequalities of land ownership . . . concentration of the ownership of draught animals, agricultural equipment and ready money intensifies the disparities. Thus we have the paradox common to many peasant societies of relatively vast inequalities in wealth amidst general poverty.' Traditionally the peasant has been described as illiterate and backward but field work in many parts of the world has shown that given the opportunity the peasant is quick to educate himself and when innovations are shown to be profitable he is swift to adopt them. The romantics maintain that peasants have a conservative and stable existence, but Franklin (1962) sees only long periods of slow evolution punctuated by sudden change. 'Archaic elements are more likely to survive within a peasant group and to remain integral parts of the culture, but it is unlikely that during the last 150 years many peasant societies have failed to experience important and perhaps significant changes, so that the study of change has become integral to the study of the modern peasantry' (Franklin, 1962, p. 4–5). Franklin calls the process responsible for much of this change '*agriculturization*'. He summarizes it thus; 'The partial incorporation of the peasantry within a market economy, the greater use of money, the appearance of usury and middlemen, the rise in rents following the increased competition for land, the weakening of communal bonds and the passing of traditional responsibilities' (1962, p. 9). In Europe this process was associated with the Industrial Revolution and is even now occurring in many underdeveloped countries. The Chinese commune and the Danish co-operative are different forms of a solution to the problem of the *paysans évolués*.

Recently Franklin (1966) has suggested that the peasant farmer should be defined on the basis of his labour commitment thus combining both social and economic aspects of previous models. The use of his own and family labour means that the peasant tends to be a satisficer giving importance to leisure as well as labour input, rather than an optimizer with regard to the output of the farm firm. Franklin (1965) uses labour as the basic differentiator in building up a model of the peasant system in relation to capitalist and socialist systems (Table 11.1). In this way regional variations and mixtures of systems can be studied and recognized on the basis of labour commitment in the various sectors of the economy (Franklin, 1965, p. 161). The concept of the peasant production system provides us with a fairly clean and noiseless model for the geographical study of peasant agriculture.

TABLE 11.1

Systems of Production

The Enterprise	*Peasant*	*Capitalist*	*Socialist*
Labour-commitment of the enterprise	Total	Non-total	Non-total
Institutional basis	Family	Family Joint stock	Combine
Control and direction	Family	Family-managerial	Managerial
Means of distribution	Barter-market	Market	Prescription-market
Media of distribution	Kind-money	Money	Money
Mechanization	Possible	Usual	Usual
Ownership (the right of . . .)			
(a) Direction	Chef d'entreprise for family	Chef d'entreprise managerial	Managerial
(b) Alienation	(1) Agnatic interdiction (2) Testamentary custom (3) Permitted	Permitted	Constitutional prohibition
Regulator	Labour supply	Market	State

Source: Franklin, 1965, p. 149.

When considering both peasant and plantation agriculture in an area, size alone may be used to discriminate between them. Edwards (1961) in Jamaica and Ooi Jin Bee (1959) in Malaya both use the figure of 25 acres as the upper limit for a peasant farm. Blaut (1961) develops a model for comparative analysis of peasant and plantation farming systems. This is based on resource materials and the amount of processing involved and takes into account the farmer's perception of his resources, his value judgements and his skill. Steward (1960) recognizes the interrelationships of peasant and plantation and states that the dispersed peasant is the counterpart of the slave plantation and the corporate peasant is the complement of the hacienda. If we consider the plantation and peasant systems as subsystems within the main system of agriculture then the concepts of general systems theory help us to understand the peasant-plantation relationship.

Environmental models

These are the basic deterministic concepts which underly much of our thinking about agricultural geography. They may be recognized at all levels. At the world scale tropical, sub-tropical and temperate areas, for example, all have been thought to have their own characteristic types of agriculture. O. E. Baker (1926), at the national level, has described American agriculture in relation to environment. In Britain we associate the wetter, western areas of Highland Britain with pastoral farming whilst the drier, eastern Lowland Britain is traditionally an area of arable farming.

These models have been widely accepted but they do tend to assume uniformity of agriculture within physically defined regions. It is often the juxtaposition of different agricultural types which is of interest to geographers. In the model of tropical agriculture (Fig. 11.2) we take as our universe that part of the world lying between the Tropics. The basic types of agriculture found within this region may be considered as sets. If we take as the basic elements of our classification firstly the types of production, arable or pastoral, and secondly the methods of production, subsistence or commercial we should then have logically four basic sets. But if we consider our model in terms of the area included in these sets then it is reasonable to combine both subsistence and commercial pastoral agriculture in one set since this type of agriculture is more common to the sub-tropics and temperate regions than to the tropics. This arrangement of sets can be shown diagrammatically by the use of a Venn diagram (Fig. 11.2).

Set A is defined as subsistence cultivation both shifting and settled. Set B is made up of commercial cultivation by both plantations and small farmers. Set C includes both nomadic and settled herding. Areas where only one of these sets is represented are becoming increasingly rare: pure subsistence cultivation is now found only in very remote places such as the upper parts of the Amazon Basin; herding with no associated cultivation is mainly found in areas of highly commercialized ranching such as northern Australia; commercial cultivation with virtually no subsistence is found in the more ad-

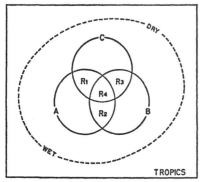

11.2 A set theory model for tropical agriculture. Set A contains subsistence cultivation; Set B commercial cultivation; and Set C herding.

vanced parts of the tropics especially where there is little room for peasants as in parts of the West Indies. By far the greater part of the tropics has a mixture of two or even all three of the basic types of agriculture. These mixtures are shown by the intersections of the sets. We can write these intersections as

$$R_1 = A \cap C$$
$$R_2 = A \cap B$$
$$R_3 = B \cap C$$
$$R_4 = A \cap B \cap C$$

R_1 is the combination of subsistence cultivation and herding, R_2 is the combination of subsistence and commercial cultivation and R_3 combines commercial cultivation and herding. R_4 covers a combination of all three basic sets and is most frequently found in conjunction with multi-racial societies. These relationships are shown in Table 11.2 (p. 435).

In most cases the intersections may be sub-divided according to the relationships between the types of agriculturist: parallel when two types of agriculture are existing side by side; dual when one group of people practise two types of agriculture; and symbiotic when two types of agriculturist have developed an interdependent system. This method of analysis does give us some idea of the complexity of agriculture in the tropics. It should be noted that as these areas become more developed the agricultural patterns become more complex.

Historical models

There are two conceptual approaches to the construction of historical models

TABLE II. 2

Tropical Agriculture

Set Intersection	Typical Environment	Sub-Division	Example
R₁	Savanna	Parallel Dual Symbiotic	Southern British Guiana Brazilian sertão Northern Nigeria – Hausa and Fulani
R₂	Humid tropics	Parallel Dual Symbiotic	Amazon – Japanese and Indian New Guinea, Yucatan West Indies
R₃	Irrigated or seasonally flooded	Parallel Dual Symbiotic	Amazon varzea Gezira Peru
R₄	Variable	—	East Africa, – European, Masai and Kikuyu North-west Argentine – European, mestizo and Amerindian

of agricultural activity. The first method looks at the problem from the point of view of the dispersal of plants and animals and culture contacts between groups of peoples. The second correlates changes in patterns of agriculture with changes in population density.

Carl O. Sauer (1952 and 1956) is an exponent of the first method. He bases his theory as to the origins of agriculture on three premises: that this new mode of life was sedentary and that it arose out of an earlier sedentary society; that planting and domestication did not start from hunger but developed in a situation in which there was both leisure and surplus food; and thirdly that primitive agriculture was located in woodlands. Sauer recognizes two ancient agricultural systems. The oldest system is known as 'hoe culture' because the main implement used is the digging stick or hoe. This method is based on mixed cropping and usually involves some bush fallowing. It is the traditional agriculture of the New World, Negro Africa and the Pacific Islands. The second basic system he identifies as a herding and sowing culture. In this system, in contrast to the hoe culture, domesticated animals play a very important role. There is little intercropping and usually only one crop season per year in the herding culture. This culture developed in the Near East and spread outwards in three directions: to the steppes of Eurasia where it became completely pastoral; to the North European plain into which the Celtic, Germanic and Slavic peoples came as cattle and horse raisers planting a few fodder crops and some rye and oats; and also along the shores of the Mediterranean. From these simple beginnings came our present complicated pattern of types of agriculture.

Malthus believed that the supply of food to the human race was inherently inelastic and this lack of elasticity was the main factor governing the rate of population growth. Thus population is seen as the dependent variable determined by preceding changes in agricultural productivity which in turn are explained by extraneous factors such as changes in technology. Ester Boserup (1965) also follows the second approach to the development of agriculture but disagrees with Malthus. She believes that population growth is the independent variable which in its turn is a major factor in determining changes in agricultural productivity. Mrs Boserup feels that her model based on changes in population density 'is conducive to a fuller understanding of the actual historical course of agriculture including the development of patterns and techniques of cultivation as well as the social structure of agrarian communities' (p. 12).

She bases her classification of systems of land use on frequency of cropping rather than on the more commonly used dichotomy of cultivated and uncultivated land, as she feels her method is more realistic when dealing with underdeveloped countries. She recognizes five types of land use as follows, in order of increasing intensity:

1 *Forest-fallow cultivation* with 20 to 25 years fallow after one or two years cultivation.
2 *Bush-fallow cultivation* with cultivation for two to as many as eight years followed by six to ten years fallow.
3 *Short-fallow cultivation* with one to two years fallow in which only wild grasses can invade the fallow land.
4 *Annual cropping.* In this system the land is left fallow for several months between the harvesting of one crop and the planting of the next. In this class may be included systems of annual rotation in which one or more of the successive crops sown is a grass or other fodder crop.
5 *Multi-cropping.* This is the most intensive system of agriculture with the same plot bearing several crops a year with little or no fallow period.

In Europe the historical development from neolithic forest-fallow to contemporary annual cropping can be recognized. In the tropics, in such countries as Nigeria where there are marked areal variations in population density, the whole range of land use systems may be seen. The most difficult step in the hierarchy is the one from long fallow to short fallow with its associated short-run decrease in output per man-hour. In the long run, however, increasing population may force changes in work habits which lead to increased productivity and increased division of labour and eventually to economic growth. Many underdeveloped countries are now at this stage of agricultural intensification and Mrs Boserup feels that her model based on past experience may help us to understand present processes.

TAXONOMIC MODELS

Classification of agricultural systems has long been a popular activity amongst geographers. Hahn (1892), Whittlesey (1936), Otremba (1950–60) and Helburn (1957) are just a few of the geographers who have worked in this field. More recently a special commission on agricultural typology has been created within the International Geographical Union. It is not intended to discuss the general problem of classification here but merely to analyse the theoretical basis of a few of the typologies set up.

Ratios and indices

In order to overcome the problem of classifying such complicated and diverse elements as farms it is necessary to develop some measure common to all. Chisholm (1964) suggests three methods of converting farm attributes to the same units based on the following indices:

1 The cash contribution of production to farm revenue
2 The cash share of inputs such as labour
3 Man/days of labour as a common index for each type of crop and livestock combination.

These indices have been much used by agricultural economists and the third index is especially useful when dealing with subsistence agriculture where few statistics are available (Clark and Haswell, 1964). There have been several other attempts to develop indices which might be used as bases for regional analysis of agriculture. One of the first of these is the distance index derived by Mather (1944). He takes farmhouses as the basic unit of his study using the formula

$$D = 1 \cdot 07 \sqrt{\frac{A}{n}}$$

where D is the average distance from one farmhouse to the nearest six farmhouses, A is the total area involved and n the total number of farmhouses. From this analysis he was able to state that farms situated to the west of the 100th meridian in the United States were more than one mile apart whilst those to the east were closer together. Livestock ratios have been used by several students to study agricultural change (Clark, 1962; Anderson, 1965). Manley and Olmstead (1965) use an index based on the monetary value of the output of the farm in a study of the geographical patterns of labour input. For studies of non-temperate agriculture Al-Maiyah (1958) uses multiple-item scaling in his study of Iraq, whilst Bhatia (1960) derives an index of crop diversification for his work in India based on the number of crops grown and the percentage of the cultivated area under different crops.

Board (1963) discusses various methods of presenting agricultural data in the form of maps of farming types and suggests the use of ratios as a basis for plotting the data. He uses Krumbein's facies maps as an analogue for his

PREDOMINANT TYPES OF FARMING IN THE EASTERN COUNTIES

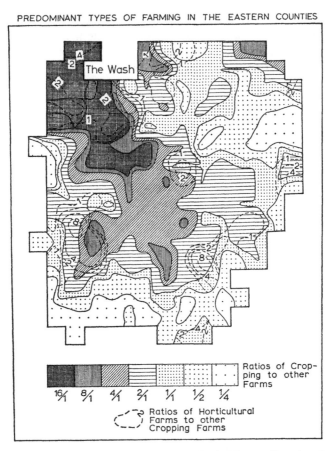

11.3 An iso-ratio map of farming types in the Eastern Counties of England (*After Board*).

cartographic presentation of farming types. Board provides us with an example based on information for Eastern England (Jackson, 1963). Using Jackson's map Board plots the ratios of cropping farms, that is farms deriving 70 per cent or more of their total gross output from crops, to all other farms (Fig. 11.3). Horticultural farms were plotted separately. The first step was to draw a square grid over the base map each square covering an area 6⅔ miles by 6⅔ miles. An arbitrary figure of 4 farms per grid square was chosen as the minimum number for which the ratio value for a square was calculated. The intervals chosen were as follows

$$\tfrac{1}{16} \quad \tfrac{1}{8} \quad \tfrac{1}{4} \quad \tfrac{1}{2} \quad \tfrac{1}{1} \quad \tfrac{2}{1} \quad \tfrac{4}{1} \quad \tfrac{8}{1} \quad \tfrac{16}{1}$$

which were converted for ease of plotting to decimals

·06   ·12   ·25   ·50   1·00   2·00   4·00   8·00   16·00

Board suggests that it is better to use proportions at the plotting stage and to employ ratios only at the final stage of compilation. The proportions appropriate to the above ratios are

·06   ·11   ·20   ·33   ·50   ·67   ·80   ·89   ·94

One could continue adding more iso-ratios to the basic pattern. The choice of ratios to be plotted and number of different ratios to be shown on the same map depends on the problem being analysed, the types of farming in the area being mapped and the aesthetic judgement of the cartographer. This method avoids abrupt boundaries on the map which have no basis in reality and gives a consistent cartographical method for comparison of different regions. It also provides a tool which can be used by other research workers with confidence.

Weaver's model

Underlying many classifications of agriculture is the idea of crop and livestock combinations. J. C. Weaver in the 1950's became interested in quantifying these combinations with a view to the study of change over time in the pattern of agricultural activity. He defined his objectives as follows, 'The central concern espoused is that of precise and objective measurement and pattern definition among the individual features and combinations of features of agricultural production as physical entities and economically active phenomena' (Weaver, 1954B, p. 286). In order to attain his objective Weaver set up a mathematical model for his crop-combination regions. He defined a theoretical curve based on the area of cropland being equally divided between the individual crops in the combination ranging from 100 per cent in a region of monoculture to 10 per cent in 10-crop-combination region. He then measured the actual crop percentage in the combination against his theoretical curve. Since he was interested in relative rank of deviation from the 'expected' not the actual magnitude of deviation, he used the standard deviation formula without extracting the square root. This was expressed as follows

$$\sigma = \frac{\Sigma d^2}{n}$$

where d was the difference between the actual and expected values of crop percentage and n was the number of crops in a given combination. The crop combination that showed the least deviation from the expected curve was recorded for every county. He also developed livestock-combination regions based on livestock units (Weaver, 1956). Weaver himself failed to develop his model further but it has been made use of by several other geographers notably Peter Scott (1957) in New Zealand, Thomas (1963) and Coppock (1964) in England and Wales and Singh (1965) in India.

The model Weaver set up has been criticized because it assumes that all crops are equal in any given crop-combination and fails to recognize those combinations in which one or more crops may be dominant although other crops are significant. Scott (1957) saw this problem, 'in any such application of the statistical procedure to a definition of agricultural regions the ranked percentage series in the basic crop-combinations should be retained however fragmented the resultant pattern' (p. 121). Weaver found it impossible to combine crop and livestock combinations because he was dealing with different units of measurement and thus areas where the interrelationships of crops and livestock are important, are not recognized. Not only do we have the loss of information in the Weaver model but we also have an element of subjectivity. Although the methods will give the same results with the same data if used by different people the original choice of the crops considered is dependent on subjective judgement. Finally the area in which Weaver developed his method, the Mid-West of the United States, was particularly suited to this type of classification. The areal unit on which Weaver based his calculations, the county, does not vary greatly in size as does the English parish for example, and thus certain computational problems were avoided. In the Middle West there is little areal differentiation based on physical variation in the landscape. Weaver was dealing with an area in which reality approached the ideal situation of the undifferentiated plain and thus combinations of crops were a fundamental basis for regional division. Other workers have not always recognized these underlying assumptions of Weaver's model. Despite its limitations Weaver's work is an important contribution to agricultural geography since he was one of the first people to attempt to set up a quantitative model for the classification of agricultural regions.

Factor analysis

The use of factor analysis as a means whereby the basic dimensions of a seemingly complex domain can be identified is well established in geographic research (Harman, 1962, p. 7). Kendall (1937) used it to derive a productivity index for crops and Hagood (1941) used it for defining regions based partly on agriculture, but geographers have only recently used this technique in the study of agriculture (Henshall and King, 1966; Henshall, 1966).

The rationale of modern factor analysis is to achieve parsimony of description. This is done by resolving a basic matrix of inter-relationships into a minimum, or at least a small number, of hypothetical variates or 'factors'. The geographer may consider these mathematical factors as similar to Hartshorne's element-complex (Hartshorne, 1960). It is assumed that the intercorrelations of the variables reflect certain underlying factors common to all the variables. The factors are made up broadly of two parts. One part is the general or common factor involved in all the variables; the other is the unique factor involved in each variable. The common factors often help to account for the maximum of the variance among the variables, whilst the unique factor indicates the extent to which correlations with other variables in the set do not account for the total unit variance of the variables. For a given matrix of correlation coefficients the position of the reference axes is indeterminate. In order to achieve a solution which has the greatest possible meaning and has consistency from analysis to analysis the reference axes are usually rotated to an orthogonal position where they are uncorrelated. The interpretation of the factors depends on the strength of the relationships between the variables and the factors as shown by the factor loadings. In addition the research worker must have a deep understanding of his problem and of the limitations of his input data if he is to identify the factors in a meaningful fashion.

One of our problems in dealing with agriculture is the wide number of variables we need to take account of and to analyse. Factor analysis, within the broad limits of the computer, can analyse the relationships between a large number of attributes or variables for many observations within a short space of time. This enables us to develop a classification based on a much larger number of variables than was possible before. Henshall and King (1966) tried to classify peasant agriculture in Barbados on the basis of crop-livestock combinations. Using factor analysis they were able to include 48 different types of crops and livestock for 150 observations or farms. The input data was in binary form based on presence or absence of individual attributes on each farm. Thus a matrix of phi coefficients was obtained as a measure of relationship between the items and then a conventional factor analysis was applied. Two approaches were adopted; the first, the R-mode analysis, focused attention on the observed correlations between the m variables measured over N cases, whilst the second, the Q-mode analysis, directed attention to the relationships between the farms as regards their attributes. Thus the factors recognized in the R-mode analysis were the basic crop-livestock combinations of the farms while those extracted in the Q-mode analysis were identified as farm types based on crop-livestock combinations. If the distribution of farms ranking high on each of the four factors recognized is mapped (Fig. 11.4) a regional pattern of crop-livestock combinations can be seen. The first factor which accounted for over 20 per cent of the total variation was recognized as a general factor loading highest on sugar cane and fruit trees and was strongly localized in the more isolated parts of the island. The second factor was identified as a cash-vegetable factor and is located fairly close to the main market in Bridgetown and the west coast hotels. The other two factors mainly recognized subsistence crops but they also produced distinctive patterns when mapped.

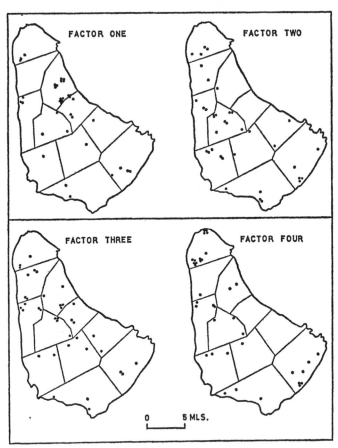

11.4 Sample farms in Barbados in the upper quartile of factor loadings for the first four factors in the Q-mode analysis.

It will be noted that the example quoted does not take into account the relative importance of individual crops or animals within the crop-livestock combination but merely their presence. However it was felt that this approach was of use in a tropical area where complicated intercropping is a feature. In addition the model set up had the built-in assumption that sugar cane was the dominant crop on virtually every farm and thus it was the secondary crops with livestock that were of interest. Of course, where the information is available, it is perfectly possible to use more detailed land use data as input for factor analysis. This method of classifying crop-livestock combinations has several advantages. A very complex combination can be handled using a computer and information is not lost in the effort to simplify the land use pattern. In addition it is possible to measure a farm's crop-livestock combination on four interval scales, these being the four factors that were extracted, and thus farms may be directly compared on the basis of their factor scores. Data collected for different areas or at different times may also be compared when analysed in a similar fashion using factor analysis.

Factor analysis has also been used to identify the relationship of a much wider range of variables to the underlying structure of agriculture and thus new agricultural regions may be recognized (Henshall, 1966). In this case the actual percentage of the holding in the various crops, livestock units and economic, social and physical characteristics of the farms were taken as attributes. Thirty-two variables were considered for 116 farms. When an R-mode analysis was carried out it was found that the 32 original variables could be collapsed into 12 new independent variates or factors. The first two factors were associated with distance variables and were identified as 'urban influence' and 'fragmentation' factors and accounted for one-third of the total variance. The next two factors accounted for one-quarter of the total variance and were identified as 'demographic' and 'motivation' factors. The situation revealed by the model showed that the basic variation in peasant agriculture is more closely related to population and settlement variables than to variations in physical environment. This is scarcely surprising in this densely populated island of gently rolling coral limestone. But if the technique is applied to other areas the order of importance of the factors may be changed or new factors revealed. Thus comparisons between areas on a common statistical basis becomes possible.

MODELS OF THE LOCATION OF AGRICULTURAL ACTIVITY

Von Thünen's model

The classical model of agricultural location outlined by von Thünen in his book *Der Isolierte Staat* (1826) is based on an econometric analysis of the estates in Mecklenburg where von Thünen farmed for forty years from 1810 until his death in 1850. Grotewold (1959) points out that only if one understands the empirical background of von Thünen's theory can one really appreciate its logic. The original model of von Thünen is inherently descriptive but later writers notably Hoover (1935), Lösch (1954), and Dunn (1954) have used the early framework as a basis for a normative model.

The underlying assumptions made by von Thünen are (1) the existence of an 'isolated state', (2) one central city as the sole market, (3) a uniform plain surrounding the city, (4) only one mode of transport, horse and cart, (5) the plain inhabited by farmers supplying the city, (6) the maximizing of profits by the farmers with automatic adjustment to the needs of the central market. However he did also consider a version of his model in which some of these assumptions of the 'isolated state' were modified by (1) introducing a navigable river on which transportation was speedier and costs only one-tenth those of land transport, (2) a minor market centre with its own trade area, (3) areal differences in the agricultural productivity of the plain around the city. Within the 'isolated state' von Thünen considered the relationship of three factors: distance of farms from the market; prices received by farmers for their goods; and finally land rent. The relationship between the first two was fairly simple. The price received by the farmer was the market price minus the cost of transportation which increased directly with distance from the market. Thus any given product was of greater value to the farmer the closer he was to the market. Land rent (Bodenrente) was defined as the return from the investment in the land. It may be derived from the expression

$$L = E(p-a) - Efk$$

where L is the locational rent per unit of land (the dependent variable), k is distance (the independent variable), E, p, f and a are constants or parameters. E is the yield per unit of land, p the market price per unit of commodity, a is the production cost per unit of commodity and f is the transport rate per unit of distance for each commodity (Dunn, 1954, p. 7). Haggett (1965, p. 161–167) discusses the substitution of a minimum-movement solution using the formula $(A/\pi)^{\frac{1}{2}}$ which simplifies the problem by omitting specific mention of the constants of market price and production cost.

The concentric rings of agricultural land use of von Thünen's hypothetical state have been well discussed by Grotewold (1959) and Chisholm (1962). The patterns seen by von Thünen 140 years ago have been radically modified by changes in transportation, new technological achievements such as refrigeration, and the replacement on the market of certain goods such as firewood by new items. In addition Lösch (1954, pp. 38–48) has pointed out that even given von Thünen's assumed conditions the formation of concentric zones was not inevitable. Yet in two spheres the nineteenth century model may still be applicable. With improvements in transportation the radius of the land use zones has become greater but the concentric zones may still be recognized on a continental scale (Backe, 1942; van Valkenburg and Held, 1952). At the other end of the scale Chisholm (1962, p. 48) considers the hamlet or farmstead in terms of von Thünen's 'isolated state' and shows that land use varies with labour input which in turn is directly related to the distance between farmhouse and field. In the less developed countries the conditions may still be similar to those of von Thünian Mecklenburg and there are several cases cited in geographical literature where land use around a settlement is directly related to distance from the settlement (for Africa by Prothero, 1957, and Steel, 1947; for India by Ahmad, 1952; for Brazil by Waibel, 1958). The von Thünen model has probably been the most fruitful in the field of agricultural geography, but there are three respects in which the model appears in need of some theoretical revision. Firstly the model is one of partial equilibrium (although Garrison and Marble, 1957 do provide a specification of the model which can be solved by simultaneous equations). Secondly the Thunian model does not take into account the influence of non-economic factors and thus is limited in its scope. Thirdly differences in scale of the central city are not considered and Harvey (1966) suggests that the failure of many medium-sized English towns to develop distinctive land use zones in the nineteenth century may be due to economies of scale experienced by the larger towns which lead to the obliterating of the smaller market centres.

Inter-regional equilibrium models

The von Thünen model operates over space and through the use of marginal analysis shows how types of land use grade into one another over a continuum although he never discusses boundary problems. Other locational models are derived by conceptualizing areas as points. Producers, factors of production and consumers are treated as located at a series of discrete points with zero transport costs between them. An analysis of comparative advantage then 'explains' differences in production at the various points. These models are of two basic types; input-output models and spatial equilibrium models (both reviewed by Harvey, 1966, pp. 365–367).

Input-output models. These models were originally devised by Leontieff as a method of analysing national economies. They have been used for the analysis of agricultural activity by Peterson and Heady (1956), Schnittkar and Heady (1958) and Carter and Heady (1959). They have tended to concentrate on the analysis of the relationships between the various regional and commodity sectors of agriculture and the effect of economic or policy changes on production patterns. They are very difficult to operate because of the wide range of input data needed.

Spatial equilibrium models. The most 'operational' technique and probably the most popular for examining the spatial equilibrium of agricultural production patterns is linear programming. Using this method and providing sufficient data is available it is possible to determine where production should be located if certain goals are to be achieved. Early work by Fox and Tauber (1955) and Judge and Wallace (1958) has been followed by many workers especially in the United States. Outstanding in this field is the work of Heady and his colleagues at Iowa State University. It is impossible to summarize all the work done as the specifications of the various models differ according to the problem being analysed. Instead one of the more recent and important works to be produced by the Iowa team will be examined in some detail.

Egbert and Heady (1964) made a study of interregional competition and the optimal spatial allocation of crop production in the United States. They state their research problem as the need 'to bring agricultural production into greater balance with "food requirements", and to cause the interregional allocation of crops to be more consistent with differential changes in technology and factor prices by regions' (p. 374). In order to do this the United States was divided into 122 producing regions and three linear programming models were used to specify which of the regions might provide the nation's requirements for wheat, feed grains, cotton and soybeans most efficiently in 1965. The first model required that soybeans be grown in rotation, the second model allowed the least-cost mix to be used, and the third model allowed continuous cropping of soybeans in each region. The models had built into them as many as 500 restraints, an upper limit on the acreage of each crop category in each region being especially important. As usual the results are only as good as the data, a fact which the authors emphasize. Non-discrete and non-linear variables could not be handled. The use of linear programming to optimize a solution places the model as a normative one but the use of empirical data gives it a strong descriptive basis. As a problem-solving technique it has many advantages over the von Thunen model.

Decision-making models

The normative models so far discussed assume perfect knowledge and rational behaviour on the part of the farmer. Man is never motivated solely by economic considerations, and social and psychological factors play an important role in determining his decisions. Recently several models have been developed which take into account these non-economic factors and their role in determining the pattern of agricultural production. Their significance in the analysis of land-use patterns in geography has been assessed by Harvey (1966, pp. 368–373).

There are two basic approaches to this problem: one is concerned with the diffusion of information and the patterns of land use produced at different stages in the acceptance of this information; the other considers the farmer's criteria for decision making in the light of his incomplete information. The models used in the first approach are those developed by workers in the field of diffusion theory and the second approach depends mainly on game-theoretical models.

Diffusion models. There are two main constraints on the flow of information through a community. The first constraint concerns the channels of communication. If these channels are constricted in some way then the flow of information will be slower than if the channels were open in every direction. These constrictions may be of several kinds: physical such as a lake or mountain range with few passes, political such as a frontier, or socio-economic for example lack of a radio or the presence of an unpopular information officer. Where these barriers are physical it has proved possible to develop a theoretical model of the expected paths of diffusion waves (Yuill, 1965). Other barriers to communication are more difficult to quantify but there is much empirical evidence to suggest that, for example, an effective agricultural officer may generate circles of decreasing acceptance of a new technique or crop around his headquarters so that his area of influence appears as an enlightened island in a sea of inertia.

The spread of information, in an age of mass communications, may rightly be regarded as less of a constraint than in the past though it is still important in underdeveloped countries (Harvey, 1966, p. 372). But various studies have shown a difference between information availability and the acceptance of the information and this problem of acceptance forms our second constraint. In the United States where mass communications are ubiquitous it has been observed that acceptance of a new technique will vary over space. Again American studies have shown that, while mass communications provide the information, personal contact is important in the final acceptance of the idea.

Much of the early work on diffusion was done in the United States (Rogers, 1962). Agricultural economists and rural sociologists were among the first to recognize the relevance of this work and the study by Bryce and Gross (1943) on the diffusion of hybrid seed corn in Iowa is a classic in this field. Hägerstrand (1953) was the first geographer to develop a model to describe the diffusion of an innovation over space. He studied the acceptance of various new agricultural practices in an area of central Sweden and showed how the innovation spread outwards from an initial centre. He designed three models to simulate the pattern of this diffusion over space, and his third model fitted the observed data remarkably well. The model contained six working assumptions:

1 Only one person possessed the information at the start.
2 The probability of the information being accepted varied through five class of 'resistance'. These classes were established entirely arbitrarily.
3 The information is spread only by telling at pairwise meetings.
4 The telling takes place only at certain times with constant time intervals.
5 At each of these times every knower tells one other person, knower or non-knower.
6 The probability of being paired with a knower depends upon geographical distance between teller and receiver of the information.

In order to make this model operational sophisticated computational techniques are needed. These are known as Monte Carlo methods. The effectiveness of the model depends on its success in simulating patterns based on empirical data. Further work in this field has been done by Wolpert (1960) but so far Sweden is the only country which has been able to supply sufficiently detailed data for long time periods to enable such models to be run.

Game-theoretical models. The basic text on game theory by von Neumann and Morgenstern only appeared in 1944 but since then the model has been shown to have wide application. Game theory was developed to deal with the problem of optimizing decisions in the face of imperfect knowledge. Although the operational procedure is complicated the basic ideas of game theory are relatively simple. In order to present these ideas of game theory let us take an example developed by Harvey (1966, p. 369): Suppose a farmer has three possible crops he can plant on his land and that he can only use one of these (mixtures being excluded). The income from these crops varies according to weather conditions of which only four are recognized. We can then construct a matrix (known as a payoff matrix) which shows the potential return from each landuse system under each weather condition.

Crop	Weather Conditions			
	1	2	3	4
A	500	550	450	600
B	600	700	300	600
C	0	2,000	0	1,000

The entries in the cells of the matrix represent the expected income level in monetary units. Given the payoff matrix the problem then becomes one of selecting criteria for the 'best' solution. Dillon and Heady (1960) suggest seven possible criteria ranging from the maximum-minimum solution, through Simon's theory of the 'satisficer', to criteria which take into account the degree of optimism or pessimism of the farmer and his gambling instincts. The criteria thus range from purely economic to purely behavioural. If we accept the maximum-minimum solution to our problem the crop A gives the highest minimum income. If we assume that all four weather conditions occur with the same relative frequency over time then the system giving the highest income in the long run is that of crop C. Crop B gives a higher income than crop A whilst avoiding the possibility of complete crop failure and consequent absence of income. Obviously the solution will depend on a considerable background knowledge of the problem on the part of the research worker (Langham, 1963, notes the importance of political pressures on farmers) in his understanding of the farmer's decisions.

If we recognize that patterns of agricultural activity are a resultant of human decisions made by a multitude of individual farm operators then it will be seen that an understanding of the processes of decision making is basic to the improvement of our models. The 'normative' theories of the economist must be considered in the light of such concepts as learning theory and the models of the behavioural scientists.

LAND POTENTIAL MODELS

Most of these models have been developed for areas of shifting cultivation in Africa. Allan (1965) gives an exhaustive theoretical survey of these areas and defines the land use factor which forms the basis of many of these models. This factor is defined (Allan, 1965, p. 30) as 'the relationship between the duration of cultivation on each of the land or soil units used in classification and the period of subsequent rest required for the restoration of fertility'. The actual acreage of land cultivated per head of population he defines as the cultivation factor.

Gourou (1962) suggests that the average density of population in Black Africa is only 25 per square mile of total area – but 750 per square mile of cultivated area. In order to define the potential density permitted by the various types of shifting agriculture observed in Africa he suggests the following formula

$$A.C/B$$

where A equals the amount of cultivable land expressed as a proportion of the whole, B equals the total length of rotation in years (period of cultivation plus period of fallow) and C equals the number of inhabitants per acre cleared annually. If we take A at 0·8 (80 per cent), B at 8 (say one year of harvest and 7 fallow or 2 years harvest and 6 years fallow) and C at 4 inhabitants per acre cleared each year then

$A.C/B = 0\cdot8 \times 4/8 = 0\cdot4$ (overall potential density measured in inhabitants per acre, or 256 to the square mile).

If such a formula were applied to the whole of Black Africa it would appear that population density could increase ten times from 25 persons per square mile to 250. Obviously the application of such a formula demands precise knowledge of the existing relationships between population and cultivable land. Regional variations in soil fertility and cultivation techniques would play an important role in determining potential patterns.

Another Belgian geographer Beguin (1964) develops several more complicated formulae for assessing land carrying potential in the Kasai province of the Congo. He defines his potential production as v, the maximum agricultural production from a given area normally obtainable under the existing agricultural system. This potential varies with the agricultural system and the physical environment. If c is the years of cultivation and j is the years of fallow then $c+j$ equals the years of the agricultural cycle. Each year part of an area changes from cultivation to fallow so a given population needs $c+j$ units of area. Each year also c units of production are harvested and the annual agricultural production equals P. The production from one unit of cultivated area is

$$\frac{P}{c+j} \text{ and therefore } v=\frac{P}{c+j}.$$

Beguin states that in Kasai yields per hectare are 0·5 tons for groundnuts, 0·3 tons for beans and 3·7 tons for cassava. Therefore in a rotation of these three crops

$$P=0·5+0·3+3·7=4·5 \text{ tons}$$

Land is cultivated for $2\frac{1}{2}$ years and lies fallow for $6\frac{1}{2}$ years.

Thus $c+j=9$ and $v=\dfrac{4·5}{9}=0·5$ tons per hectare.

Potential for an agricultural system

In most tropical agricultural systems the number of crops is far greater than in the first model and there are usually several interlocking rotations. So for his second model Beguin includes multiple rotations. Then the model may be stated as follows

$$v=\frac{\sum_{i=1}^{l} v_i . r_i}{\sum_{i=1}^{l} r_i}$$

where l is the number of rotations, v_i is the potential of the ith rotation and r_i is the area occupied by rotation i. When this formula is applied to data from Kasai where there are six principal rotations the results are as follows

$$v_i=0·5, \ 0·47, \ 0·3, \ 0·3, \ 0·42, \ 0·42 \text{ tons per hectare}$$

when weighted by the area assigned to each rotation

$$v=\frac{(0·5)32+(0·47)12+(0·3)25+(0·3)6+(0·42)7+(0·42)12}{32+12+25+6+7+12}$$

$$=0·41 \text{ tons per hectare}$$
$$\text{or}=41 \text{ tons per square kilometre}$$

Population density potential

This varies according to the standard of living of the population and Beguin's third model takes this into account. If u is the unit of production per capita and if the population desire one ton each per year then they would need $\dfrac{u}{v}$ land units. The area (s) necessary for a given population (p) so that each inhabitant could obtain a certain annual production (u) practising a certain type of agriculture on a given soil (v) would be

$$s=\frac{pu}{v}$$

Then the maximum density (d) of this population would be

$$d=\frac{p}{s}=\frac{v}{u}$$

For example if $v=60$ tons per square kilometre in an area where there are 1,000 people producing one ton each then

$$s=\frac{pu}{v}=\frac{1,000 \cdot 1}{60}=16\cdot67 \text{ square kilometres}$$

If $s=20Km^2$ then $p=\dfrac{20\times60}{1}=1,200$ people.

If $u=1\cdot2$ tons then $d=\dfrac{60}{1\cdot2}=50$ inhabitants per square kilometre.

Areal differentiation

It is more realistic to assume that the potential varies through area in response to environmental and cultural differences. If we accept that the maximum (v_M) and minimum potential (v_m) are on a continuum and vary linearly then mean potential ($\bar{v}$) equals $\dfrac{v_M+v_m}{2}$.

Then in a region of 100 square kilometres with a potential varying linearly from 100 tons per square kilometre to 60 tons per square kilometre, giving one ton per inhabitant per year the population maximum would be

$$p=\frac{100\times\dfrac{100+60}{2}}{1}=8,000 \text{ inhabitants}$$

Thus the maximum population density is 80 inhabitants per square kilometre.

Beguin develops several more refined formulae which take into account sequent occupation of land in which later occupants are forced on to land of lower potential; non-homogeneous regions; input of work per capita; and change over time as it affects population, production per capita and potential. Thus the model is both dynamic and wide ranging as to its input. Its main disadvantage is the difficulty of obtaining the empirical data on which it is based. Beguin is one of those rare people who are capable of building theoretical models on a strong basis of fieldwork. This deductive approach to model building may provide us with our most useful models for the development of agricultural geography.

FUTURE TRENDS

Perhaps the basic problem of agricultural geography is that of aggregation. This has two aspects, of scale and depth. Firstly we have a brick-laying problem, that of combining micro-studies of our basic building block, the farm, into statements about macro-areas. Secondly we have the problem of combining data concerning the physical environment with information related to the human environment.

The development of new equipment and techniques is enabling us to solve some of these long-standing problems. Modern digital computers have a vast storage capacity and can make large numbers of calculations in a very short space of time. Thus they help us to deal with the problem of combining information from many farms for an area and they make it possible to apply simulation models with their need for many iterations, to behavioural data. New statistical methods of combining quantitative and qualitative data make it possible to improve our studies in depth.

New methods are enabling us to make greater use of our information and forcing us to re-examine some of our theoretical concepts. The need for a new approach to agricultural geography has been seen (Reeds, 1964; Brookfield, 1964) and it is hoped that the models now available to us will help to bring this about.

REFERENCES

AHMAD, E., [1952], Rural Settlement Types in the Uttar Pradesh (United Provinces of Agra and Oudh); *Annals of the Association of American Geographers*, 42 (3), 223–246.

ALLAN, W., [1965], *The African Husbandman*, (London), 505 pp.

AL-MAIYAH, ALI MOHAMMED, [1958], *An Analysis of the Spatial Relationships Among Agricultural Phenomena in Iraq*, 1953; Unpublished Ph.D. dissertation, State University of Iowa, USA.

ANDERSON, J., [1965], The use of fodder and livestock units in agricultural geography: A Case Study of Soviet Land Use Policy; *Annals of the Association of American Geographers*, 55 (4), 603.

BAKER, O. E., [1926], Agricultural Regions of North America; *Economic Geography*, 2, 459–493.

BEAL, G. M. and ROGERS, E. M., [1960], The Adoption of Two Farm Practices in a Central Iowa Community; *Special Report*, No. 26, *Agricultural and Home Economics Experiment Station*, (Ames, Iowa).

BEELEY, B. W., [1965], Agricultural Change: A 'Field' Study; *Annals of the Association of American Geographers*, 55 (4), 605.

BEGUIN, H., [1964], *Modèles géographiques pour l'espace rural africain*, (Brussels).

BERTALANFFY, L. VON, [1951], An outline of general system theory; *British Journal of the Philosophy of Science*, 1, 134–165.

BHATIA, S. S., [1960], An Index of Crop Diversification; *Professional Geographer*, XII (2), 3–4.

BIRCH, J. W., [1954], Observations on the Delimitation of Farming Type Regions, with special reference to the Isle of Man; *Transactions of the Institute of British Geographers*, 20, 141–158.

BLAUT, J. M., [1953], The Economic Geography of a One-Acre Farm in Singapore: A Study in Applied Microgeography; *Malayan Journal of Tropical Geography*, 1, 37–48.

BLAUT, J. M., [1959], Microgeographic Sampling: A Quantitative Approach to Regional Agricultural Geography; *Economic Geography*, 35 (1), 79–88.

BLAUT, J. M., [1961], The Ecology of Tropical Farming Systems; *Revista Geografica*, 28 (1), 47–67.

BOARD, C., [1963], Some Methods of Mapping Farm Type Areas; Unpublished Paper.

BOSERUP, E., [1965], *The Conditions of Agricultural Growth*, (London), 124 pp.

BOWDEN, L. W., [1965], The Diffusion of the Decision to Irrigate; *University of Chicago, Department of Geography, Research Paper* No. 97.

BRANDNER, L. and KEARL, B., [1964], Evaluation for Congruence as a Factor in the Adoption Rate of Innovations; *Rural Sociology*, 29, 288–303.

BROOKFIELD, H. C., [1962], Local Study and Comparative Method: An Example from Central New Guinea; *Annals of the Association of American Geographers*, 52 (3), 242–253.

BROOKFIELD, H. C., [1964], Questions on the Human Frontiers of Geography; *Economic Geography*, 40 (4), 283–303.

BUCHANAN, R. O., [1938], A Note on Labour Requirements in Plantation Agriculture; *Geography*, 23, 156–164.

BUCHANAN, R. O., [1959], Some Reflections on Agricultural Geography; *Geography*, 44, 1–13.

BUNGE, W., [1962], Theoretical Geography; *Lund Studies in Geography, Series C, General and Mathematical Geography*, No. 1, 210 pp.

BUTLER, J. B., [1960], *Profit and Purpose in Farming*; (*A Study of Farms and Smallholdings in part of the North Riding*), *Department of Economics, University of Leeds*, 68 pp.

CARTER, H. C., [1963], Representative Farms – Guides for Decision Making; *Journal of Farm Economics*, 45 (5), 1449–1455.

CHISHOLM, M., [1962], *Rural Settlement and Land Use*, (London), 207 pp.

CHISHOLM, M., [1964], Problems in the Classification and use of the Farming Type region; *Transactions of the Institute of British Geographers*, 35, 91–103.

CLARK, A. H., [1962], The Sheep/Swine Ratio as a Guide to a Century's Change in the Livestock Geography of Nova Scotia; *Economic Geography*, 38 (1), 38–55.

CLARK, C. and HASWELL, M. R., [1964], *The Economics of Subsistence Agriculture*, (London), 218 pp.

COPPOCK, J. T., [1964A], Crop-livestock and enterprise combinations in England and Wales; *Economic Geography*, 40 (1), 65–81.

COPPOCK, J. T., [1964B], *Agricultural Atlas of England and Wales*, (London), 255 pp.

DAY, L. M., [1963], Use of Representative Farms in Studies of Interregional Competition and Production Response; *Journal of Farm Economics*, 45 (5), 1438–1444.

DEAN, G. W. and BENEDICTIS, DE M., [1964], A Model of Economic Development for Peasant Farms in Southern Italy; *Journal of Farm Economics*, 46 (2), 295–312.

DILLON, J. L. and HEADY, E. O., [1960], Theories of Choice in Relation to Farmer Decisions; *Agricultural and Home Economics Experiment Station, Iowa State University, Research Bulletin* 485, (Ames, Iowa).

DUNN, E. S., [1954], *The location of agricultural production*, (Gainesville), 115 pp.

EDWARDS, D., [1961], *An Economic Study of Small Farming in Jamaica*, (Mona, Jamaica), 370 pp.

EDWARDS, D. and REES, A. M. M., [1964], The Agricultural Economist and Peasant Farming in Tropical Conditions; In *International Explorations of Agricultural Economics*, (Ames, Iowa), 73–85.

EDUCATION, MINISTRY OF, [1960], Geography and Education; *Ministry of Education Pamphlet* No. 39, (London).

EGBERT, A. C., HEADY, E. O. and BROKKEN, R. F., [1964], Regional Changes in Grain Production; *Iowa Agricultural Experimental Station Research Bulletin*, No. 521, (Ames, Iowa).

ELLIOT, F. F., [1928], The 'Representative Firm' Idea Applied to Research and Extension in Agricultural Economics; *Journal of Farm Economics*, 10, 481–489.

EDMONSON, M. S., [1960], Hybrid Corn and the Economics of Innovation; *Science*, 132: 3422, 275–280.

FIRTH, R., [1951], *Elements of Social Organization*, (London).

FOX, K. and TAUBER, R., [1955], Spatial Equilibrium Models of the Livestock Feed Economy; *American Economic Review*, 45, 584–608.

FRANKLIN, S. H., [1962], Reflections on the Peasantry; *Pacific Viewpoint*, 3 (1), 1–26.

FRANKLIN, S. H., [1965], Systems of Production: Systems of Appropriation; *Pacific Viewpoint*, 6 (2), 145–166.

FRANKLIN, S. H., [1966], Personal Communication.

GARRISON, W. L. and MARBLE, D. F., [1957], The Spatial Structure of Agricultural Activities; *Annals of the Association of American Geographers*, 47, 137–144.

GERLING. W., *Die Plantage*, (Würzburg).

GOULD, P. R., [1963], Man against his environment: a game-theoretic framework; *Annals of the Association of American Geographers*, 53, 290–297.

GOUROU, P., [1962], *Agriculture in the African Tropics: the Observations of a Geographer*; Paper read at the University of Oxford.

GREGOR, H. F., [1965], The Changing Plantation; *Annals of the Association of American Geographers*, 55 (2), 221–238.

GROTEWALD, A., [1959], Von Thünen in Retrospect; *Economic Geography*, 35 (4), 346–355.

HÄGERSTRAND, T., [1952], The propogation of Innovation waves; *Lund Studies in Geography, Series B, Human Geography*, 4, 3–9.

HÄGERSTRAND, T., [1953], *Innovationsförloppet ur korologisk synpunkt*, (Lund).

HAGGETT, P., [1965], *Locational Analysis in Human Geography*, (London), 339 pp.

HAGOOD, M. J., [1943], Statistical methods for delineation of regions applied to data on agriculture and population; *Social Forces*, 21, 288–297.

HAHN, E., [1882], Die Wirtschaftsformen der Erde; *Petermann's Mitteilungen*, 38. 8–12.

HARVEY, D. W., [1966], Theoretical concepts and the analysis of agricultural land-use patterns in geography; *Annals of the Association of American Geographers*, 56, 361–374.

HARTSHORNE, R. and DICKEN, P., [1935], A classification of the Agricultural Regions of Europe and North America on a Uniform Statistical Basis; *Annals of the Association of American Geographers*, 25, 99–120.

HEADY, E. O. and EGBERT, A. C., [1964], Regional Programming of Efficient Agricultural Production Patterns; *Econometrica*, 32 (3), 374–386.

HELBURN, N., [1957], The bases for a classification of World Agriculture; *Professional Geographer*, 9, 2–7.

HELLER, C. F., [1964], The use of Model Farms in Agricultural Geography; *Professional Geographer*, 16 (4), 20–23.

HENSHALL, J. D., [1964], *The Spatial Structure of Barbadian Peasant Agriculture*; Unpublished M.Sc. Thesis, McGill University, Montreal.

HENSHALL, J. D. and KING, L. J., [1966], Some Structural Characteristics of Peasant Agriculture in Barbados; *Economic Geography*, 42 (1), 74–84.

HENSHALL, J. D., [1966], The Demographic factor in the structure of Agriculture in Barbados; *Transactions of the Institute of British Geographers*, 38, 183–195.

HOOVER, E. M., [1936], The measurement of industrial localization; *Review of Economics and Statistics*, 18, 162–171.

HUTCHINSON, H. W., [1959], Comments on E. T. Thompson's 'The Plantation as a Social System'; In *Plantation Systems of the New World*, Social Science Monography No. VII, 37–40, (Washington).

INNIS, D. Q., [1961], The Efficiency of Jamaican Peasant Land Use; *The Canadian Geographer*, V (2), 19–23.

JACKSON, B. G. et al., [1963], The Pattern of farming in the Eastern Counties; *Occasional Papers No. 8, Farm Economics Branch, School of Agriculture, Cambridge*.

JONASSON, O., [1925–26], Agricultural Regions of Europe; *Economic Geography*, 1, 277–344 and 2, 19–48.

JOHNSON, R. W. M., [1964], The Labour Economy of the Reserves; *Department of Economics, Occasional Paper No. 4, University College of Rhodesia and Nyasaland*, (Salisbury).

JONES, C. F., [1928–30], Agricultural Regions of South America; *Economic Geography*, 4, 1–30, 159–186 and 267–294; 5, 109–140, 277–307 and 390–421; 6, 1–36.

JONES, W. D., [1930], Ratio and isopleth maps in Regional Investigation of Agricultural Land Occupance; *Annals of the Association of American Geographers*, 20, 177–195.

JUDGE, G. G. and WALLACE, T. D., [1958], Estimation of spatial price equilibrium models; *Journal of Farm Economics*, 40, 801–820.

KENDALL, M. G., [1939], Geographical Distribution of Crop Productivity in England; *Journal of the Royal Statistical Society*, 102, 21–62.

KEUNING, H. J., [1964], Agrarische Geografie: Doelstelling, Ontwikkeling, Methoden; *Tijdschrift van het Koninklijk Nederlandsch Aardrijkskundig Genootschap* (Amsterdam), 81 (1), 10–19.

KRUMBEIN., W. C., [1956], Regional and local components in facies maps; *Bulletin of the American Association of Petroleum Geologists*, 40, 2163–2194.

LANHAM, W. J. and COUTU, A. J., [1964], Area Resource Adjustments for Specified Net Revenue Goals and Levels of Factor Prices on Farms in Economic Area 7, N. Carolina; *Agricultural Economics Information Series No. 109, Department of Agricultural Economics, North Carolina State of the University of North Carolina at Raleigh.*

LANGHAM, M. R., [1963], Game theory applied to a policy problem of rice farmers; *Journal of Farm Economics*, 45 (1), 151–162.

LATHAM, J. P., [1959], The Distance Relations and some other characteristics of Cropland Areas in Pennsylvania; *Technical Report No 6, Contract NONR 551 (O1) University of Pennsylvania.*

LEONTIEFF, W. W., [1953], *Studies in the structure of the American Economy*, (New York).

LÖSCH, A., [1954], *The Economics of Location*; (translated by W. W. Woglom), (New Haven).

MANLEY, V. P. and OLMSTEAD, C. W., [1965], Geographical Patterns of Labour Input as related to output indexes of scale of operation in American Agriculture; *Annals of the Association of American Geographers*, 55 (4), 629–630.

MARTIN, A., [1958], *Economics and Agriculture*, (London), 169 pp.

MATHER, E. C., [1944], A linear-distance map of farm population in the United States: *Annals of the Association of American Geographers*, 34, 173–180.

MCCARTY, H. H., [1954], Agricultural Geography; In *American Geography: Inventory and Prospect*, James, P. E. and Jones, C. F. (Eds.), (Syracuse), 258–277.

MCCLURE, J. A., [1964], The use of correlation and factor analytic techniques in comparing farmland potentials; *Annals of the Association of American Geographers*, 54 (3), 430.

MIKHEYEVA, V. S., [1963], An economic-mathematical model of the Location of Farm Production by Regions of the Soviet Union; *Soviet Geography, Review and Translation*, 4 (3), 24–29.

MILLER, B. R. and KING, R. A., [1964], Models for measuring the impact of technological change on location of marketing facilities; *Agricultural Economics Information Series No. 115. Department of Agricultural Economics, North Carolina State of the University of N. Carolina, Raleigh.*

NEUMANN, J. VON and MORGENSTERN, O., [1944], *Theory of games and economic behaviour*, (Princeton).

OOI, JIN BEE, [1959], *Land, People and Economy in Malaya*, (London).

OTREMBA, E., [1950–60], Allgemeine Agrar-und Industriegeographie; *Erde und Weltwirtschaft*, 3, 213–229 and 343–351.

PADILLA, E., [1960], Contemporary Social-Rural Types in the Caribbean Region; In *Caribbean Studies: A Symposium*, V. Rubin (Ed.), (Seattle), 22–28.

PICO, R., [1959], Comments on 'Some observations relating to population dynamics in plantation areas of the New World' by T. Lynn Smith in *Plantation Systems of the New World*. Social Science Monographs No. VII, (Washington).

PLATT, R. S., [1930], Pattern of Occupancy in the Mexican Laguna District; *Transactions of the Illinois State Academy of Science*, 22, 533–541.

PLATT, R. S., [1942], *Latin America: Countrysides and United Regions*, (New York), 564 pp.

PLAXICO, J. S. and TWEETEN, L. G., [1963], Representative farms for Policy and Projection Research; *Journal of Farm Economics*, 45 (5), 1458–1465.

PROTHERO, R. M., [1957], Land Use at Soba, Zaria Province, Northern Nigeria; *Economic Geography*, 33, 72–86.

REDFIELD, R., [1956], *Peasant Society and Culture*, (Chicago), 92 pp.

REEDS, L. G., [1964], Agricultural Geography: Progress and Prospects; *Canadian Geographer*, 8 (2), 51–63.

ROGERS, E. M., [1962], *Diffusion of Innovations*, (New York).

ROGERS, E. M., [1964], Bibliography of Research in the Diffusion of Innovations: *Research on the Diffusion of Innovations No 1, Department of Communication, Michigan State University*.

RYAN, B. and GROSS, N. C., [1943], The Diffusion of Hybrid Seed Corn in Two Iowa Communities; *Rural Sociology*, 8, 15–24.

SAUER, C. O., [1952], Agricultural Origins and Dispersals; *American Geographical Society, Bowman Memorial Lectures*, 2.

SAUER, C. O., [1956], The Agency of man on the Earth; In *Man's Role in Changing the Face of the Earth*, Thomas, W. L. Jr., (Ed.), (Chicago).

SCOTT, P., [1957], The Agricultural Regions of Tasmania; *Economic Geography*, 33, 109–121.

SINGE, H., [1965], Crop combination regions in the Malwa tract of Punjab; *Deccan Geographer*, 3 (1), 21–30.

SMITH, T. L., [1959], Some observations relating to population dynamics in plantation areas of the New World; In *Plantation Systems of the New World*, Social Science Monographs, VII, (Washington), 126–132.

STAMP, L. D., [1948], *The Land of Britain: Its Use and Misuse*, (London), 507 pp.

STEEL, R. W., FORTES, M. and ADY, P., [1947], Ashanti Survey 1945–6: An Experiment in Social Research; *Geographical Journal*, 110, 149–179.

STEWARD, J. H., [1960], Perspectives on Plantations; *Revista Geografica*, No. 52, 26 (1), 77–85.

TAKAYAMA, T. and JUDGE, G. G., [1964], An interregional activity analysis model for the agricultural sector; *Journal of Farm Economics*, 46 (2), 349–365.

TAYLOR, G., [1930], Agricultural Regions of Australia; *Economic Geography*, 6, 109–134 and 213–242.

THOMAS, D., [1963], *Agriculture in Wales during the Napoleonic Wars: A study in the geographical interpretation of historical sources*; (Cardiff).

THÜNEN, J. H. VON, [1875], *Der Isolierte Staat in Beziehung auf Landwirtschaft und*

Nationalokonomie; Third edition, (Berlin). (A first edition of Part I appeared in 1826).

VALKENBURG, S. VAN, [1931–36], Agricultural Regions of Asia; *Economic Geography*, 7, 217–237; 8, 109–133; 9, 1–18, 109–135; 10, 14–34; 11, 227–246, 325–337; 12, 27–44, 231–249.

VALKENBURG, S. VAN and HELD, C. C., [1952], *Europe*, (New York).

WAIBEL, L., [1958], *Capítulos de geografia tropical e do Brasil*, (Rio de Janeiro).

WEAVER, J. C., [1954], Changing Patterns of cropland use in the Middle West; *Economic Geography*, 30 (1), 1–47.

WEAVER, J. C., [1954], Crop-combination regions in the Middle West; *Geographical Review*, 44 (2), 175–200.

WEAVER, J. C., [1954], Crop-combinations regions for 1919 and 1929 in the Middle West; *Geographical Review*, 44 (4), 560–572.

WEAVER, J. C., [1954], Isotope and Compound: A framework for Agricultural Geography; *Annals of the Association of American Geographers*, 44 (3), 286–288.

WEAVER, J. C., HVAG, L. P. and FENTON, B. L., [1956], Livestock Units and combination regions in the Middle West; *Economic Geography*, 32, 237–259.

WEAVER, J. C., [1956], The county as a spatial average in agricultural geography; *Geographical Review*, 46 (4), 536–565.

WHITTLESEY, D., [1936], Major agricultural regions of the Earth; *Annals of the Association of American Geographers*, 26, 199–240.

WOLF, E. R., [1955], Types of Latin American Peasantry: A Preliminary Discussion; *American Anthropologist*, 57 (3), 452–471.

WOLPERT, J., [1963], *Decision making in Middle Sweden's Farming: A Spatial Behavioural Analysis*; University of Wisconsin Ph.D., University Microfilms, (Ann Arbor).

WOLPERT, J., [1964], The decision process in spatial context; *Annals of the Association of American Geographers*, 54 (4), 537–558.

YUILL, R. S., [1965], A simulation study of barrier effects in spatial diffusion problems; *Michigan Inter-University Community of Mathematical Geographers, Discussion Papers*, 5.

ZABKO-POTOPOWICZ, A., [1957], The development of the geography of agriculture since World War I; *Przeglad geograficzny*, (Warsaw), 29 (1), 21–46.

Index

D. R. STODDART

Since this book is primarily about ideas and people rather than places, no attempt has been made to index the many incidental references to place-names in the text. The intention of this index is, first, to locate references to people and their writings, and second, to trace the main ideas common to many of the papers. Page-references to literature citations are given in italics, thus: *236*.